URBAN POLICY IN PRACTICE

This text is a detailed, up-to-date account of urban policy in Britain. With a focus on local government, it combines a study of policy in practice with a critical assessment of recent developments.

Using a wide range of examples, the author shows how policies and services are responding to today's urban problems, stressing in particular the need for corporate strategies, democratic control and sustainable development. The broader debates about urban change and the local state are related to new directions in policy and management.

The opening chapter discusses urban policy as a framework for allocating resources to meet needs. How to define 'need' and how to assess it are major themes of the book. Subsequent chapters explore how policy is put into practice using objective setting, performance indicators and evaluations of its effects.

The book includes a clear and detailed study of local government finance, and an analysis of the growing role of research in urban policy and public administration.

The author argues for an urban policy which is 'democratic, strategic and enables the management of services close to the consumer'.

Tim Blackman is Head of Research at Newcastle City Council and Chair of the Local Authorities Research and Intelligence Association.

URBAN POLICY IN PRACTICE

Tim Blackman

London and New York

First published 1995
by Routledge
11 New Fetter Lane, London EC4P 4EE

Simultaneously published in the USA and Canada
by Routledge
29 West 35th Street, New York, NY 10001

Typeset in Garamond by Solidus (Bristol) Limited

Printed and bound in Great Britain by
Biddles Ltd, Guildford and King's Lynn

British Library Cataloguing in Publication Data
A catalogue record for this book is available from the British Library

Library of Congress Cataloging in Publication Data
Blackman, Tim.
Urban policy in practice / Tim Blackman.
p. cm.
Includes bibliographical references and index.
1. Urban policy—Great Britain. 2. Municipal government—Great Britain. I. Title.
HT133.B54 1995
307.76'0941—dc20 94-9428

ISBN 0–415–09299–X
0–415–09300–7 (pbk)

CONTENTS

FIGURES

TABLES

PREFACE

In writing this book I have sought to bring together an academic interest in urban policy with the experience of working in a large metropolitan local authority and putting policy into practice. The book draws on both the academic and professional literature for examples and analysis. However, I have subjected this literature to my own 'litmus test': does it make sense in terms of real-life policy-making and implementation? The result, I hope, is an account which will give students an academic knowledge of the field. But my intention is to add to this the type of practical understanding which students will need if they move into careers in the public sector, or in the growing number of private companies with professionals working on public sector contracts.

The book reflects three particular interests. The first is an emphasis on *corporate* policy, reflecting my own position as the head of a research unit involved with all the departments and many of the professions of a metropolitan council, as well as parts of the local public sector. I have therefore focused on a series of corporate policy areas: the overall scope and purposes of urban policy, strategic approaches to making and implementing policy, need assessment and the allocation of resources, business planning in the public sector and assuring the quality of services, community development, the carrying out and use of research, the improvement of education and training, the development of local economies, public health, and sustainable development. These areas cut across professional and disciplinary boundaries. I believe this perspective reflects an emerging trend being brought about by an emphasis on outcomes – such as healthy cities and sustainable economic growth – rather than just inputs to urban policy. Such an interdisciplinary approach will not be new to social scientists working in the academic sector.

The second interest arises from my role as a local authority research manager. Not only does the book reflect my belief that policy-makers and service managers need to make greater use of research, but it also reflects my conversion to management! Policy-making and implementation have to be managed. It makes little sense to teach about policy without showing how the process is managed to deliver results. Again, contemporary trends seem to be moving towards an emphasis on both research and management. However, I am opposed to the 'new managerialism' which has been imposed on many parts of the public sector to the cost of democratic control and public participation.

Third, the book is based on a commitment to local government. Unfortunately, support for local government is not a growing trend in the United Kingdom, in contrast to most other European countries. Since 1979, many powers and functions have been removed from local authorities in a process that has made the UK one of the most centralised countries in Europe. There needs to be a major redistribution of this centralised power back to sub-national government, both to local councils and preferably to a system of regional government. I hope that readers will consider that the book makes a case for this.

ACKNOWLEDGEMENTS

Many people have supported me in writing this book. Roberta Woods has not only contributed her expertise and experience as an academic and local councillor, but has given me the love and inspiration to see me through writing it. Our daughter, Maeve, has also put up with the stress and strain, and I hope seeing her dad's evening keyboard bashing in print will be an education if not a reward.

I have benefited so much from various discussions with the following people in Newcastle City Council that they should be named as having knowingly or unknowingly contributed to this book: Geoff Norris, Paul Keenan, Tim Hibbert, Chris Stephens, Graham Armstrong, Carolyn Stephenson and Dave Johnstone. Carol Fitz-Gibbon, John Harvey, Barbara Wallace and Paul Vittles were important influences. I am also grateful to Dave Byrne, John Mohan and Derek Birrell for their support and encouragement during the time that I have worked both in higher education and local government. Tristan Palmer, Jane Mayger, Caroline Cautley and Alan Fidler at Routledge provided invaluable support and guidance as the book moved through its various stages from an original idea some two years ago.

Of course, responsibility for what is in this book lies entirely with me.

I am grateful to the following for permission to reproduce copyright material: Cambridge University Press for Figure 2.1 (Titterton, 1992); Frank Cass & Company for Figure 5.1 (Hancox, Worrall and Pay, 1989) and Figure 7.2 (Blackman, 1993); Combat Poverty Agency, Dublin, for Table 8.1 (Donnison *et al.*, 1991); Mike Coombes (1993) for Table 6.1; Department of City and Regional Planning, University of Wales College of Cardiff for Figure 2.2 (Hambleton, 1992); Department of Health (1993a) for Tables 7.1, 7.2, 7.3, 7.4 and Figure 7.1; Department of the Environment (1991 and 1994) for Tables 3.2 and 4.7; Gateshead Health Authority for Table 10.1 (Henley, 1991); the Controller of Her Majesty's Stationery Office for Tables 2.1, 4.3, 4.4 and 4.5 (Audit Commission, 1993a), Tables 11.2 and 11.3 (Hansard, 1992c) and Figure 5.2 (PIEDA, 1992); *Inside Housing* for Table 4.10; P. M. Jackson and B. Palmer (1992) for Figure 5.3; Leicester City Council for Appendices 1 and 2; Local Government Management Board for Table 9.1

(Brooke, 1990); *Local Government News* for Figure 9.1 (Sheldon, 1992); Longman Group Limited for Table 10.3, Figure 10.1 (Byrne *et al.*, 1985) and Figure 9.2 (Marvin, 1992); *Public Finance* and the author, T. Fairclough, for Table 4.6; The Royal Statistical Society for Table 7.7 (Marsh *et al.*, 1991); Social Information Systems (1991) and Warwickshire County Council for Table 5.4; *Town & Country Planning* for Table 9.2 (Rawcliffe and Roberts, 1991); York City Council for Table 5.1.

Tim Blackman
May 1994

ABBREVIATIONS

AEF	aggregate external finance
AMA	Association for Metropolitan Authorities
ANC	African National Congress
ASB	Aggregated Schools Budget
BCA	Basic Credit Approval
BCGN	Bristol Community Groups Network
BTEC	Business & Technician Education Council
CATs	City Action Teams
CBA	cost–benefit analysis
CCT	compulsory competitive tendering
CDP	Community Development Project
CEMR	Council of European Municipalities and Regions
CFCs	chlorofluorocarbons
CFF	Common Funding Formula
CIPFA	Chartered Institute of Public Finance and Accountancy
COSLA	Confederation of Scottish Local Authorities
CTSS	council tax for standard spending
DoE	Department of the Environment
EAGGF	European Agricultural Guidance and Guarantee Fund
EIB	European Investment Bank
EMG	Ethnic Minority Grants
ERDF	European Regional Development Fund
ESF	European Social Fund
ESG	education support grant
ET	Employment Training
EU	European Union
FMI	Financial Management Initiative
GAE	grant-aided expenditure
GDP	gross domestic product
GIS	geographical information systems
GM	grant-maintained
GNI	generalised needs index

GNVQ	General National Vocational Qualification
GREA	grant-related expenditure assessment
HIPs	housing investment programmes
HMOs	houses in multiple occupation
HRAs	housing revenue accounts
HRG	Housing Research Group
ILF	Independent Living Fund
LARIA	Local Authorities Research and Intelligence Association
LDDC	London Docklands Development Corporation
LEA	local education authority
LEC	local enterprise company
LIF	Local Initiative Fund
LMS	local management of schools
MoHs	medical officers of health
MSC	Manpower Services Commission
NRAs	neighbourhood renewal areas
NVQ	National Vocational Qualification
OFSTED	Office for Standards in Education
OPCS	Office of Population Censuses and Surveys
OR	operational research
PPBS	planning programme budgeting systems
PSB	potential schools budget
PSBR	Public Sector Borrowing Requirement
QA	quality assurance
RSG	Revenue Support Grant
SATs	Standard Assessment Tasks
SCAs	Supplementary Credit Approvals
SLAs	service level agreements
SRB	Single Regeneration Budget
SSAs	standard spending assessments
SSRG	Social Services Research Group
STG	special transitional grant
SVQ	Scottish Vocational Qualification
TEC	training and enterprise council
TQM	total quality management
TSS	Total Standard Spending
UDC	urban development corporation
UKHFAN	UK Health for All Network
WHO	World Health Organisation
WRFE	Work-Related Further Education
ZBB	zero-based budgeting

Part I

URBAN POLICY AND THE STATE

1

THE SCOPE AND PURPOSES OF URBAN POLICY

This book is about how urban policy in the United Kingdom is put into practice. It considers the roles of central government departments, executive agencies appointed by central government, and elected local authorities. Since the most substantial agencies in the local public sector are still local authorities, much of the book is based on local government material. However, the decline in the importance of local authorities that has occurred during the last fifteen years makes it essential to consider the new machinery of central government which has developed to implement policy in the UK's regions and cities.

This growth of centralisation has been one of the most dominant and controversial features of urban policy in recent years. The role of local government has been strongly challenged by legislation since 1979. This has transferred functions from elected local authorities to new executive agencies, limited the powers of local authorities to spend and develop their own policies, and substituted private sector providers of services for local council workforces. The power relations between central and local government have had a potent influence on urban policy, making it an important focus for political as well as urban studies.

This chapter begins by explaining the book's rationale and structure. This is followed by a section on the evolution of urban policy, from its origins in nineteenth-century industrialisation, through its expansionist post-war period, to the 1980s and 1990s, when public services have been under pressure to retreat back to a residualist welfare role. The next section considers how urban policy has responded to urban change. The recent reforms of public services are then introduced, particularly the strategies of privatisation and promoting competition. These have given rise to the idea of public sector agencies being 'enablers' which assess needs but do not necessarily provide services directly themselves. Instead, within the limits of their budgets, they purchase services from competing providers according to the needs and priorities they have established on behalf of their local populations.

The book has four parts, each comprising chapters linked to a particular theme. The theme of Part I is urban policy and the state. This begins with the present chapter which defines urban policy, considers how it has developed over time, and introduces key aspects of the nature of urban policy in the UK today. Chapter 2 reviews urban policy as it relates to local government and Chapter 3 as it relates to central government. Chapter 4 then describes how urban services are financed, including new methods for funding local authority services introduced in April 1994, the role of local taxation, and specific aspects of the financing of schools, care in the community and social housing.

The theme of Part II is general approaches to urban policy. Three particular approaches have been chosen because of their recent growth in importance as *corporate policy concerns* about the quality of services, community participation and how needs are identified. Chapter 5 explains how the quality of public services is managed and how the outcomes of policies are evaluated. It includes discussion about some of the key controversies in these areas. Chapter 6 considers community development, explaining approaches to supporting people who are disempowered by poverty or discrimination to organise and act collectively to express their needs and improve conditions. Chapter 7 discusses the uses and value of applied research about the nature and distribution of social needs, the quality of services and the outcomes of policy.

Part III has major policy goals as its theme. Three key goals are considered because of their importance to any definition of urban well-being. Chapter 8 discusses goals for education, training and local economic development, linking these three areas together because of the vital significance of knowledge and skills to the UK's economic future. Chapter 9 considers the fundamental goal of safeguarding the sustainability of urban society, discussing the role of urban policy in sustainable development and realising environmental objectives. Finally, the prevention of illness and promotion of health are essential to everyone's quality of life; Chapter 10 considers the important role that urban policy has in public health.

Parts II and III, therefore, present a view of the major purposes of urban policy in the UK today and explain how these purposes can be realised in practice. Best practice requires a knowledge of how to assure quality in public services, how to involve local communities and pursue community development, and how to undertake and use research to improve services and policies. The key goals which the book identifies for urban policy – education, training and economic development; sustainability and environmental quality; and the safeguarding and promotion of health – are fundamental, not only to an urban policy that can have positive effects and public support but also to any vision of a modern and caring society.

Part IV comprises the book's concluding chapter. This discusses how urban policy addresses 'needs' and how this ideal is hindered by the pressing need for democratic reform in the UK.

Two major issues are absent from this brief summary of each chapter. The first is the question of social inequality which is at the centre of much public policy debate. The concerted attack on unemployment, poverty and social segregation necessary to achieve a more just society is largely a question for national government policy and, increasingly, action at a European and international level. It demands a strategy for full employment and a much more progressive taxation and benefits system. Holman (1993), for example, has recently pointed out that in the UK if the wealthiest 20 per cent lost a fifth of their disposable income, the incomes of the poorest 20 per cent could be doubled. Many urban local authorities target their services on areas and groups in greatest need, but their ability to tackle the causes of these problems is very limited.

The second major issue is equal opportunities and anti-discriminatory practice. This is a very important area of work for many urban local authorities. It also requires legislative action at national level, with many people now arguing that the UK should have a bill of rights to give stronger force to the protection of civil liberties and outlawing of discrimination. The specialist treatment of equal opportunities is beyond the scope of this book, but it is hoped that readers will find this principle reflected in the material that follows.

The book's definition of urban policy is a wide one. Urban policy is essentially about the welfare of local residents in an urban society. This involves planning and delivering public services and supporting the development of the local economy. A number of agencies in the local public sector have a role in urban policy, but the most important are elected local authorities. This is not only because local authorities provide or purchase on behalf of local residents a large range of services, but also because they have a local democratic mandate to represent the interests of the local population in all contexts, regional and national. Where they do not have direct responsibilities, such as health care or income maintenance, they often seek to influence the government agencies that do.

Local authorities and other public sector agencies often pursue policies of prioritising particular groups who, because of economic disadvantage or discrimination, are exposed to poverty, unemployment or social exclusion to a much greater extent than the rest of the population. Common priorities are children in low-income families, people who are elderly and frail, people who have disabilities, are unemployed, lack skills or are educationally disadvantaged, and minority ethnic groups. Urban policy frequently seeks to prioritise in this way.

Many cities and towns in the UK have experienced a decay in traditional inner city communities and the development of large housing estates with multiple disadvantage on the urban periphery. Poverty has persisted despite periods of economic growth. It is in these areas that urban policy is most relevant, providing public services and increasing access to employment and training. Unfortunately, this is all the harder in the face of national government policies in recent years which have caused a widening of inequalities in British society. But even with less regressive policies economic growth cannot be relied upon to 'lift all boats' (Donnison *et al.*, 1991). Urban policies are needed to create equal opportunities to benefit from education, employment and public services, to live in pleasant and safe environments with adequate shops and amenities, and to participate in community and civic affairs.

The planning and implementation of urban policy consumes a large proportion of public expenditure. It therefore needs to be justified in terms of positive effects that can be demonstrated and have public support. Policy-making should be a rational process which implements desired goals, such as targeting training and job creation on high unemployment areas or improving the quality of public services. Increasingly, it is expected that there are measurable targets and arrangements for monitoring and adjusting actions which are not proving to be effective. However, opinion often differs about what the goals of policy and the appropriate means for realising these goals should be. This is why, in principle, policy is under democratic control, although the reality often does not accord with this ideal.

Among the reasons for this 'democratic deficit' are the effects of factors beyond the control of politicians, the many defects of democratic institutions, and the fact that politicians are advised by professional officers who often have considerable influence. Indeed, the role of representative politics in determining public policies has been considerably weakened by the growth of 'technocratic government' and the major influence of powerful unelected corporate bodies on policy-making and implementation, especially large multinational companies (Dunleavy and O'Leary, 1987).

Although the values behind policy goals will always be issues for argument, once these goals are established the process of implementing policy needs to be as rational as possible, informed by monitoring and evaluation. There is also a growing expectation that implementation is 'user-oriented', with consumers and the wider public actively involved in defining the quality and value of the services they receive.

THE EVOLUTION OF URBAN POLICY

In order to understand the role of urban policy in today's society it is helpful to review the origins of state intervention in urban areas.

The unprecedented increase in population which accompanied industrialisation in nineteenth-century Britain, Western Europe generally, and in North America saw a dramatic growth of towns and cities. At the beginning of the nineteenth century only one-fifth of the British population lived in towns; by the end of the century this had increased to four-fifths. Migration to urban areas created large working-class areas very rapidly. For example, in Birmingham the working-class area of Ladywood expanded from 10,000 in 1841 to over 40,000 just thirty years later. The pattern was repeated all over the country as people sought industrial employment.

Urbanisation occurred at a pace that outstripped the ability of pre-industrial political and social institutions to maintain social order. These had been based on networks of clients, kin and friends, and centred on the country residences of the aristocracy. Early attempts to manage this new urban society through *ad hoc* bodies undertaking specific tasks such as cleansing streets, providing watchmen and relieving destitution failed to cope with the scale of urbanisation and with growing demands for local self-government. Poverty and overcrowding were sources of ill-health, unrest and poor labour productivity. Lash and Urry comment that British cities:

> grew up with very little planning, before the growth of nationally organized professional experts, of sanitary engineers, medical inspectors of health, civil engineers, social workers, town planners, and so on. Builders, interested in short-term profit maximization, constructed as many dwellings as possible on each acre of building land, which were not conducive to effective communications and other infrastructural services.
>
> (Lash and Urry, 1987: 95)

Of particular concern to the powerful and wealthy of nineteenth-century Britain was the maintenance of order and authority as traditional society broke up with urbanisation. Bauman (1987) describes how traditional football matches came to be branded as dangerous to public health and order, while in Derby the establishment of a police force in 1835 was accompanied by the instruction that, 'Persons standing or loitering on the footway without sufficient cause, so as to prevent the free passage of such a footway ... may be apprehended and taken before a magistrate' (Delves, quoted in Bauman, 1987: 65). Bauman concludes that:

> The destruction of pre-modern popular culture was the main factor responsible for the new demand for expert administrators, teachers, and 'social' scientists.
>
> (Bauman, 1987: 67)

Concern with an unruly 'underclass' in the new towns and cities was met with both philanthropy and resistance to extending state assistance to the poor, which would burden the taxpayer. The themes of an urban underclass and the tax burden of providing public services survive in debate about urban policy today (Macnicol, 1987).

During the nineteenth century, various legislation was passed by the British parliament which constructed a complex patchwork of local authorities and *ad hoc* boards across the country with new responsibilities for public health, roads, housing, relief of the poor and education. This structure began to be simplified and reformed towards the end of the century. In England and Wales, a system of elected county councils and county borough councils in an upper tier of local government, and district and parish councils in a lower tier, was established by 1899. In Scotland, a two-tier structure of city councils and county councils in the upper tier, and burghs and district councils in the lower tier, was finally created in 1929.

The establishment of local government was part of a period of transition from the 1880s to the 1920s during which Britain became a mass democracy. Democracy was to see increasingly successful political action by the Liberal Party and by the emergent Labour Party to establish a welfare state in Britain. Local government was a key part of the welfare state and underwent major growth as public services expanded during the 'post-war boom'.

The expansion of the welfare state which took place during the period of post-war economic growth up to the 1973–5 world slump saw public expenditure consume an increasingly large share of Britain's gross domestic product (GDP). Local government spending paralleled this growth, as did health service expenditure. In 1955 local authorities spent 9 per cent of the country's GDP; by 1975 this had increased to 15 per cent, when their expenditure was almost three times greater in real terms than in 1955. More people were employed to run and deliver services and many public building projects were undertaken. This expansion brought about major improvements in public services:

> The period 1955–75 saw bigger, airier and lighter schools, better trained teachers, and more carefully planned books and syllabuses. Larger, more elaborate leisure centres were provided, as well as better staffed and stocked libraries. Better roads, pedestrian lights and public buildings were developed. This list indicates only the tip of the iceberg in the improvements introduced during this period.
>
> (Stoker, 1991: 7)

During much of this growth, the basic structure of local government remained largely unchanged until pressures for reorganisation began to mount. Reform to create bigger and more effectively managed local councils began in London in the early 1960s and in the rest of Britain in the early 1970s. During 1972–4 two tiers of new local authorities were created in England and Wales: in the upper tier there were 53 county councils (6 metropolitan and 47 non-metropolitan) and in the lower tier 369 district councils (36 metropolitan and 333 non-metropolitan). In Scotland, the new upper tier consisted of 9 regional councils and the lower tier of 53 district councils. Benington (1975) described the reform as turning local government into 'big business'; by European standards very large authorities were created and new management structures were introduced modelled on corporate management ideas derived from big US firms and state departments.

Cockburn (1977) argued that the preoccupation with budgets and plans was an attempt to contain growth in public expenditure, whilst national economic policy sought to achieve economic growth in the private sector as the post-war boom slowed. By the mid-1970s, economic problems were pressing and expansion of the welfare state began to falter. Inner city poverty was threatening disorder and militancy. Government policy targeted deprivation in the inner cities, but Cockburn's interpretation is that urban policy became concerned with the direct management of people as well as resources, using the techniques of public participation and community work to 'incorporate' protest when public expenditure cuts began.

Inner city problems of growing poverty and ageing populations were part of a wider trend: a reversal of the major population movements into the industrial cities. In the 1950s firms and population began to move out of the major conurbations, particularly the inner urban areas. Over the following decades many urban manufacturing firms contracted or closed down altogether. Economic and population growth shifted to smaller towns and some rural areas, notably in the south of England, leaving the inner cities in decline and their populations increasingly dependent on public services and social security benefits. Social polarisation became an intensifying feature of urban areas, often marked by racism.

Social polarisation has had an increasingly strong housing tenure aspect. Government financial and fiscal policies since 1980 have reduced council housing to a tenure of last resort for people who for reasons of unemployment, low pay, illness, old age or lone parenthood cannot afford to buy a house and are dependent upon housing benefit to meet all or part of their rent (Forrest and Murie, 1990). The result is that a map of urban deprivation will today largely mirror the distribution of council housing (see Figures 1.1 and 1.2). Indeed, there has been a distinct trend in recent years towards many local government services being limited to a welfare role for the poor, rather than community services for everyone.

Local authorities have been proactive in the face of these problems and a series of policy responses emanated from local government, despite budgetary constraints, during the 1980s. Local economic development expanded as a new and non-statutory area of activity, eventually recognised but also constrained by legislation in 1989 (see Chapter 8). Urban authorities were faced with massive job losses in heavy industry and responded with new policies for cities as places of commercial, social and cultural life. This was also a period of intense political conflict between some local authorities and central government over spending controls and issues such as subsidising public transport (Duncan and Goodwin, 1988).

Today, there is less conflict but local government has become a weaker player in a local public sector which includes a variety of government agencies operating at local level beyond the influence of local electorates, from district health authorities to training and enterprise councils. Unemployment is a nation-wide problem although still most severe in the old

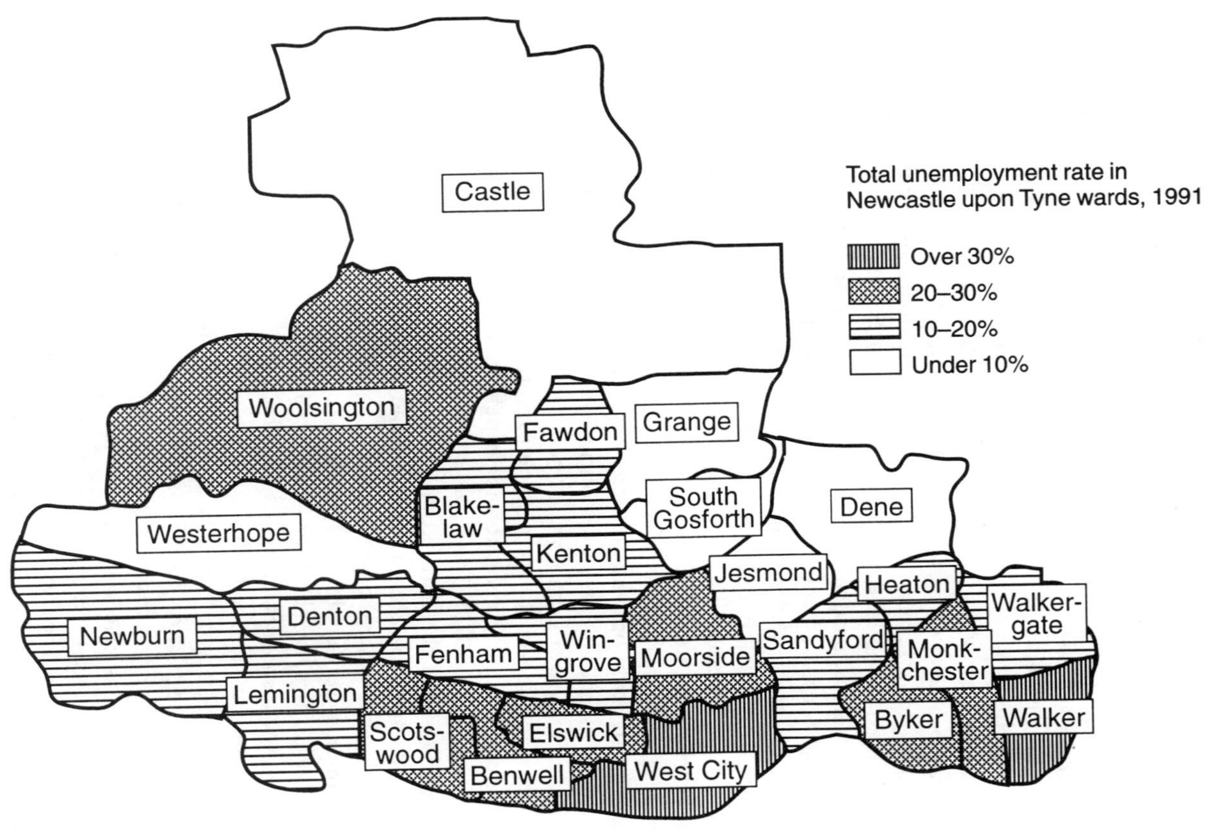

Figure 1.1 Unemployment in Newcastle upon Tyne
Source: 1991 Census.

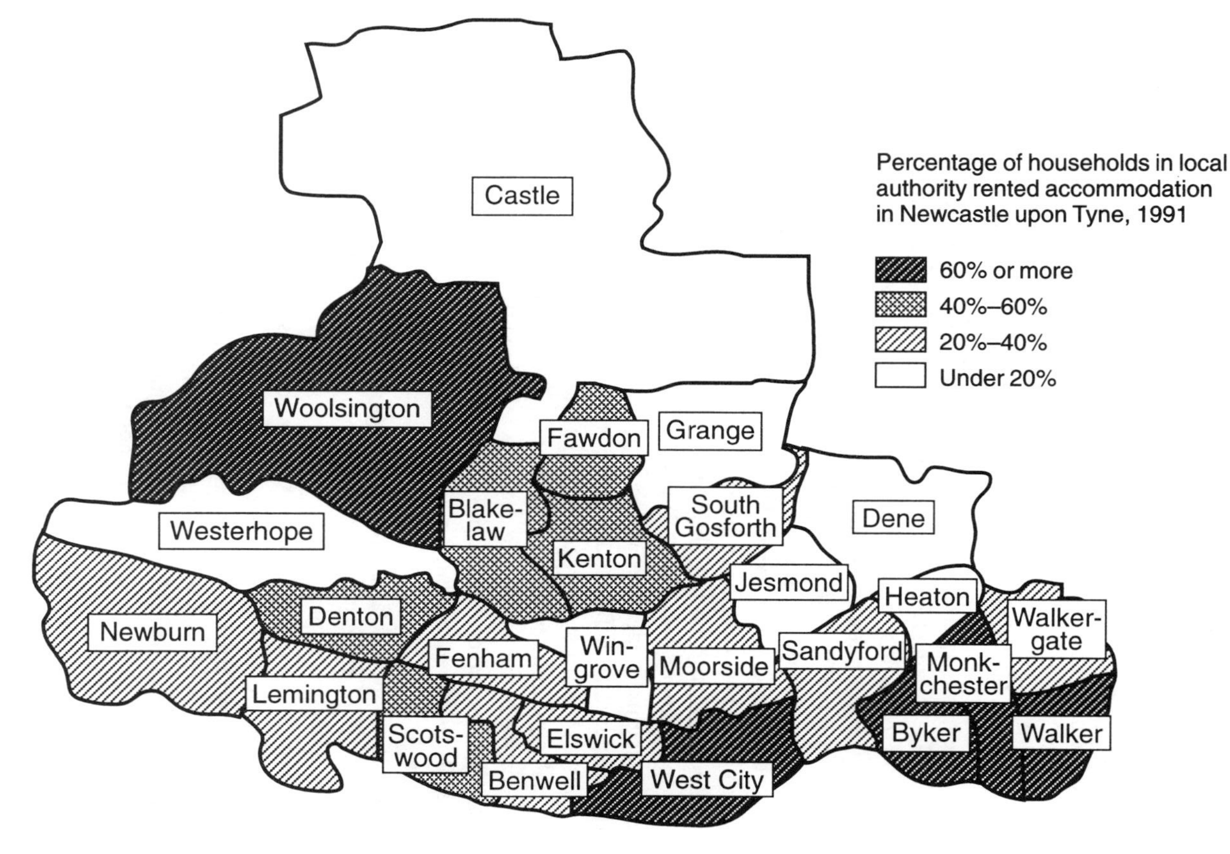

Figure 1.2 Households in local authority housing, Newcastle upon Tyne
Source: 1991 Census.

industrial areas. A new paradigm of public sector management based on contracts and competition now complements policies that have brought about the dismantling of the old welfare state. In its place is an increasingly two-tier society based on the private consumption of housing and services for those able to afford it, and the provision of cash-limited public services – often subject to an individual assessment of eligibility and a means-test – for those who cannot.

URBAN POLICY IN AN URBAN SOCIETY

Urban policy as a general term is about the activities of government in urban areas. The Department of the Environment (1993a) defines an urban area as a built-up area with a minimum population of approximately 1,000 people. However, it is common to include in the definition of urban areas both cities and their commuter hinterlands. These hinterlands often encompass large areas of countryside which are still essentially part of an urban society, looking to towns and cities for services and employment.

Urban areas are partitioned by the boundaries of their local councils and other public sector bodies, but such administrative units are only one definition of towns and cities. Tyneside, for example, has a distinctive and widely recognised identity as the conurbation along the River Tyne in North-East England. But today there is no 'Tyneside' local authority. The area consists of four metropolitan district councils, comprising different towns, villages, industrial and commercial areas, with large commuter flows into them from the bordering counties of Northumberland and Durham. These counties have also received emigrants from the conurbation in a process of *counter-urbanisation*, and Tyneside itself is declining in population.

Given such inter-relationships, 'urban' is a term which describes the general nature of society in Britain and other advanced industrial countries, rather than particular places. Even in the less densely populated 'rural' areas society is often essentially 'urban'. One of the most visible signs of this is the physical infrastructure of urban societies, in particular the level of development of transportation and communication infrastructure, and the mobility and mass culture it supports. This development reflects the needs of the economic base of urban society, which is dominated by the large manufacturing and service industries of modern capitalism. In such societies, 'the urban process implies the creation of a material physical infrastructure for production, circulation, exchange and consumption' (Harvey, 1981: 103).

Some urban sociologists have argued that the city should be defined as a site of 'collective consumption' – centres which provide urban public services and where 'urban social movements' can form to struggle for more and better public services (Castells, 1979; Dunleavy, 1980; Saunders, 1986). Although cities have long had this role, today the identity of cities is much more as places of mass fashion and entertainment. They have become dominated by a private sector 'retail culture' and what Gardner and Sheppard (1989) call its 'new cathedral': the shopping centre. This has increasingly replaced their traditional roles as industrial centres and public places (Harvey, 1989; Lash and Urry, 1987).

The transition to post-industrial land uses in British towns and cities has been controversial. Byrne (1989) argues that many deindustrialised sites should be retained as areas zoned for manufacturing industry in the event of industrial recovery, rather than have public subsidy spent on redeveloping these areas for private housing and shopping, which occurred on a wide scale in many urban areas during the 1980s. Others have argued that the challenge is to attract any investment to such redundant sites. Thus, Stoker and Young write:

> The argument was that if policy-makers succeeded in levering private sector investment onto sites like these, this would stimulate market forces into operating. If this could be done, private sector investment would become an attractive proposition again . . . However, entrepreneurial planning is not just about recreating the market. It has the wider goal of getting people and vitality – even a sense of festival – back into neglected areas. It aims to inject them with life and excitement, so that there is a buzz, even on Sunday evenings . . .
>
> (Stoker and Young, 1993: 38)

The complex means by which 'entrepreneurial planning' has achieved this transition for many sites is described by Stoker and Young. As is discussed in Chapter 3, the problem with these projects has often been their failure to relate the redevelopment of land to local needs for employment and public services. More fundamentally, the decline of the UK's manufacturing base which these projects signify is of major concern in terms of the country's economic future.

The decline of manufacturing industries and the growth of unemployment and low-pay employment sectors such as retailing and leisure have combined with the residential segregation of low-cost rented housing in Britain to produce 'deprived areas' in towns and cities. Deprived areas are where people live with problems of poverty, unemployment, crime, youth disturbances, ill health and environmental damage. The underlying economic problems of these areas are caused by forces beyond the locality, particularly the investment, disinvestment and rationalisation decisions of the multinational companies which dominate the UK economy. Nevertheless, urban policy can achieve meaningful improvements in quality of life, and the local public sector itself is a major employer with significant economic and social impacts. However, as Wolman and Goldsmith (1992) argue, it cannot be *assumed* that urban policy necessarily increases urban well-being. Its impact should always be analysed critically and the programmes which result from urban policy initiatives monitored and evaluated in terms of benefits and costs. Indeed, urban policy cannot be separated from politics and the whole question of who wins and who loses as a result of policy decisions.

Some public services are 'pro-rich' – the top income group use them more than the bottom – while others are 'pro-poor'. Bramley and Le Grand (1992) carried out an interesting analysis of survey results from Cheshire County Council and found that education, libraries and parks were pro-rich services, while council housing and social services were pro-poor. Overall, the use of local services was found to be income-neutral. Pro-poor services are usually those where access is subject to an individual assessment of need, such as council housing and social services. Pro-rich services are usually those that people choose to use, such as staying on at school or using a library. In response, many councils have adopted policies of targeting their resources and services. This can consist of either providing an enhanced level of services to 'priority areas' which score highly on census deprivation indicators for example, or targeting subsidies on people with low incomes or in receipt of means-tested benefits by offering reduced prices for using leisure facilities and services such as home care. To be effective, it is important that such targeting actually encourages and increases the uptake of services among target groups.

Although it is often argued that public services should be about providing a better quality of life for people with low incomes, the need for many services is governed by other factors as well. The need for local authority social services such as home helps, for example, has been found to be largely independent of income or social class, depending much more on the level of disability, the availability of support from the family and the local authority's policy on charges for services (Bramley and Le Grand, 1992). And although poverty and deprivation are marked features of life in many urban areas, it is also the case that race, gender, disability and locality are causes of significant inequalities in access to services and jobs, independent of income.

As well as targeting present needs, urban policy must also anticipate change. Demographic, economic and social changes have to be monitored and met by developing local strategies to meet future needs. In the UK, marked changes have occurred during the last ten years. The number of secondary school pupils, for example, has fallen by almost 25 per cent, while the number of people aged 85 or older has increased by more than 50 per cent. As a result, many schools have 'surplus places' which need to be rationalised to make most effective use of public spending whilst taking into account any likely future increase in numbers. Care services for older people have had to expand, but again in a way that balances scare resources with levels and types of need. Many areas continue to experience deindustrialisation and disinvestment, causing a series of problems from unemployment and dereliction, to crime and ill-health. A combination of these problems can cause rapid neighbourhood decline. An ability to monitor key 'early

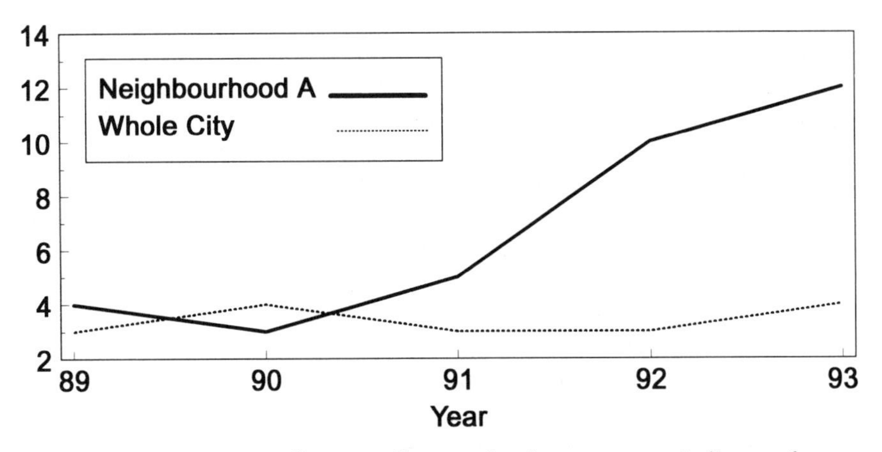

Figure 1.3 Long-standing council tenants leaving an area – an indicator of neighbourhood deterioration in Newcastle upon Tyne (tenants over five years as a percentage of all terminations*)
Note: *Excludes deaths and succession.

warning' indicators such as people leaving the area can help public agencies to deliver preventive measures at an early stage, such as improvements to housing and security (see Figure 1.3).

The complexity of urban change means that it cannot be managed within a single organisation. Organisations have to work together and in partnership with the local community. Stoker and Young (1993) argue that the management of change and complexity is best achieved through 'networks' of organisations and individuals working together, focusing on the same problems and pooling resources. Local government provides an obvious framework for this corporate approach because it is already multifunctional and is increasingly geared to working in alliances with other parts of the local public sector. As subsequent chapters illustrate, there is growing evidence of networking establishing itself as an important means of making and implementing urban policy.

THE NEW PUBLIC SECTOR MANAGEMENT

The local public sector has typically consisted of organisations with hierarchical departments and professionally led services. In local government, electoral checks have been fairly weak given that they involve at the most one annual election covering many issues. In such a situation, people's needs may become secondary to *producer* interests: professional officers seeking to maximise their budgets and local politicians serving sectional interests. This type of analysis has figured strongly in Conservative Party ideology, and Conservative governments have responded by introducing competition, 'that is, replace monolithic state services with numerous competing providers' (Hambleton, 1992: 5). Market signals then provide feedback about what service users want (and can afford) rather than elections. Hospitals, schools, refuse collection services, managers of council housing and leisure services, and providers of caring services are all now expected to compete for custom. The service is usually secured on behalf of local consumers by a 'purchasing authority'. These include the housing and social services departments of local authorities, government agencies such as the Department of Employment, district health authorities and fundholding general practitioners. These purchasers are expected to assess needs and establish priorities before spending cash-limited budgets. Consultation with users and consumer research are expected to form an important part of this needs assessment.

The separation of strategy – including the specifying and monitoring of contracts for services – from the day-to-day running of services is the major justification for the 'purchaser–provider split' that has been widely introduced into the British public sector. This innovation is partly a response to the very real managerial problems that had been experienced by corporate managers and policy-makers in large public sector bureaucracies during the 1970s–80s. Corporate managers such as chief executives sought centralised control over functionally specialised departments with their own professional group interests. Hoggett sums up how this led to centralised units losing grip of strategic control and becoming embroiled in departmental issues:

> Increasingly ... the centre of bureaucracy was forced to adopt a policing role which led it towards progressive encroachment on routine and administrative operations and details themselves. Hence the constant experience of policy makers within state bureaucracies in the later 1970s and 1980s of being bogged down in detail, of never having time to focus upon real strategic issues.
>
> (Hoggett, 1991: 247)

The introduction of a split between organisations which purchase public services and organisations which provide them, mediated by a contract, marks a profound break with the past dominance of single bureaucracies planning and providing services. It has seen the end of local government's monopoly over the provision of many social services and the management of schools and careers services, social housing and publicly funded recreational facilities. In many cases, local authorities' own services have to compete with private firms for contracts.

Compulsory competitive tendering (CCT) has forced many councils to separate the roles of service specification and monitoring from that of service delivery. Legislation has extended CCT to a wide range of services in recent years with the purpose of introducing competition into the public sector. Thus, the purchaser–provider split has not just been a management reform;

it has been part of a strategy of competition and privatisation.

Separating out commissioning and providing functions in the public sector has intrinsic merits that have not just appealed to right-wing politicians. The National Commission on Education (1993), which was otherwise critical of Conservative education policy, recently advocated the establishment of new local education and training boards as 'purchasers' of education to replace local education authorities. The Commission proposed that the boards contract with schools, further education colleges and private training companies on an arm's-length purchaser–provider basis. It was argued that this arrangement would secure the advantages of devolved management whilst strengthening strategic planning and quality assurance. There are, however, problems in the NHS where the purchaser–provider split has seen a substantial growth in bureaucracy to manage contracts and inflexibilities arising from the way contracts prescribe what can and cannot be done. In local government, the split has also been experienced as having advantages and disadvantages (see Chapter 3).

Although their service-providing role is increasingly subject to competition or undertaken by contractors, local councils retain many overall responsibilities for assessing needs and planning services. In recent years this has seen the emergence of topic-based plans for meeting needs with a focus which is wider than the council's own direct services. Examples include community care plans, local housing strategies and economic development statements, as well as more traditional documents such as land use and development plans (Carter *et al.*, 1992). Because these plans encompass many organisations in the public, private and voluntary sectors, they have a key role as a focus for consultation between organisations. They also have key roles in linking policies to the resources likely to be available, and in specifying explicit goals and targets.

With direct services subject to competition, the role of local authorities in ensuring that these services are provided has given rise to the notion of the 'enabling council'. In a parliamentary debate on local government reform, the then Secretary of State for the Environment, Michael Heseltine, defined this notion in the following terms:

> Many of the Government's policies have been designed to move power to individual members of the community – in particular, the right to buy, the tenants' charter and local management of schools. Compulsory competitive tendering has also been part of that process . . . The success of a local authority must be measured not by its size, but by the quality of the services delivered, from whatever source, by its responsiveness to its clients and customers and by the value that it squeezes out of each pound of taxpayers' money.
>
> (*Hansard*, 1991, col. 402)

CONCLUSION

The changes that have occurred in urban policy in recent years have been driven by the ideology of Conservative governments. Some changes have appeared to be responses to real problems in the public sector, such as how to bring new investment to huge tracts of disused land in inner cities, how to enhance the responsiveness of public services to their consumers, or how to improve the management of services by devolving responsibility for decisions. All the major political parties have acknowledged problems in these areas. Indeed, 'citizens' charters' were first initiated by Labour local authorities, decentralisation has been widely pursued by Liberal Democrat and Labour councils, and it was a Labour government which appointed the Taylor Committee in 1975 which recommended devolving more powers from local education authorities to schools. There is broad political consensus about the need continually to strive for improvements in the accountability of the public services, in how they are managed and in effective working with the private sector. Where there has not been consensus is the introduction during the 1980s and 1990s of competition and privatisation on a large scale, and the exceptionally rapid pace of change which has been forced on managers and public sector workers. The rationale for this has been strongly ideological, accompanied by a rhetoric of free enterprise and individual responsibility. Beyond this rhetoric is a more serious debate about the scope and purposes of public services and state intervention generally, a debate which will become increasingly important if economic growth and taxation do not keep pace with growing social needs (see Chapter 4).

The rapid pace of change is not just a result of new legislation. The rate of change in the economy and society generally has accelerated. The era of mass production, mass consumption and the traditional nuclear family has passed. Society is much more differentiated. If there is one term which sums up the contemporary challenge of putting urban policy into practice, it is the management of change and complexity.

2

URBAN POLICY AND LOCAL GOVERNMENT

This chapter is about the changing role of local government in urban policy. The major theme is the reduction in the powers and status of local authorities. This has occurred despite the case for a strategic elected body at local level. Local government has a unique role in making corporate policies which are affirmed by the local electorate and cut across the separate responsibilities of individual services, both those of local authorities and other organisations. This is discussed in the next section of the chapter, with a number of examples. Subsequent sections discuss the effects of competition and individualism on local government, and consider the new models to which 1990s local councils are expected to aspire: the 'enabling council' and the 'competitive council'. The decline in local government bureaucracy which these terms signify is then examined in terms of the effects of underlying economic change on the public sector. The chapter concludes by relating the general decline of the welfare state to the emergence of a residualised public sector in a more unequal and unstable society.

In a unitary state such as the UK, local government is subject to the sovereignty of parliament. Parliament legislates for local authorities to undertake certain specific functions only. In this capacity they are education authorities, housing authorities, planning authorities and road authorities; they run social services departments, act as local environmental and consumer protection agencies, fund public leisure facilities, and provide economic development services. Their specific functions, however, contrast with the *general competence* that local authorities have in a number of other European countries where, within certain safeguards, they can undertake any activities that meet the needs and problems of their areas (Batley and Stoker, 1991). The sovereignty of parliament is one reason why the UK refrained from signing the European Charter of Local Self Government in 1985.

Most European countries are moving towards a greater devolution of state power to regions and local authorities (see Chapter 11). However, although British local authorities are unusually restricted by central government prescriptions regarding their functions, taxation powers and ability to decide levels of spending on local services, they are much larger than continental

Table 2.1 The size of local government expenditure: European comparisons

	Net expenditure as % of GDP (1987)
United Kingdom	11.6
Denmark	29.6
Netherlands	17.6
Germany	5.7
France	5.7*
Italy	5.0
Portugal	4.0

Note: *France: municipalities only (9.3% including Regions and Depts).
Source: Audit Commission, 1993a.

municipalities. They still employ large numbers of professional and manual staff, and have substantial executive responsibilities.

Local councils in Britain are key actors in urban policy. Their spending accounts for about 25 per cent of public expenditure and they employ around 10 per cent of the country's total workforce. In many areas, the local council is the largest single employer, investor, purchaser, landowner and estate agent. Net expenditure by UK local government represented 12 per cent of national gross domestic product in 1987. As Table 2.1 shows, this is relatively large by European standards. The smaller expenditure in Germany, France and Italy reflects the existence of regional authorities with major responsibilities, a tier of government which is absent in the UK.

Several central government departments have policy responsibility for services that are organised and run by local authorities. However, the Department of the Environment (DoE) in England and the Scottish, Welsh and Northern Ireland Offices in the rest of the UK have a central responsibility for policy towards local government. This includes the structure and areas of local authorities, their general powers and rules of procedure, and their financial basis and resources. In England, regional offices of the DoE play a key role in ensuring that local authorities fully understand policies and procedures, and that they are properly implemented at the local level.

Local councils are Britain's democratic local governing authorities at sub-national level. Local government is the only way of providing 'comprehensively, democratically and accountably for the social and environmental needs of identifiable communities' (Alexander, 1991: 64). However, the significant reduction in this role in recent years, brought about by central government regulating local authority spending, limiting local taxation and removing functions, has led to a debate about whether local government is being reduced to the local administration of central government policies (see Stoker, 1991; Batley and Stoker, 1991). This is certainly the trend, but local councils still retain an important autonomous role in localities. Exactly what localities should form the basis for local government has since 1992 been the subject of a fundamental review by the Local Government Commission in England and the Scottish and Welsh Offices. To date, a series of radical and controversial proposals for change has resulted from this review, including the abolition of many local authorities within present two-tier structures and their replacement with new unitary authorities with populations in the range of 150–250,000.

Although local authorities are still large organisations with major responsibilities, their importance in urban policy has declined due to the loss of powers and functions. This has been part of a shift away from elected local government towards unelected local administration in the UK. Many activities previously carried out by local councils are now undertaken by housing associations, housing action trusts, funding bodies for schools, urban development corporations, and training and enterprise councils. In only one area, the funding of care services in the community, have local authorities seen a significant addition to their functions since 1979. Research by *The Guardian* newspaper in 1993 concluded that by 1996 Britain will have more than 7,700 public bodies controlled wholly or partly by central government appointees, a quadrupling of the number that existed in 1979 (Beavis and Nicholson, 1993). They will be responsible for spending nearly £54 billion, almost a quarter of total public expenditure.

Conservative government ministers have argued that the creation of executive agencies has focused the work of public services by creating single-purpose bodies, brought people with business skills into their management, and strengthened the consumer's voice by treating the public as customers. Critics claim that their effect is to fragment public services so that it is difficult to co-ordinate housing, education, economic development and other functions. They argue that they have removed accountability to local electorates and that they are run by supporters of Conservative government policies in areas and regions which have not voted for these policies.

For key services that remain with local government, legislation has introduced a split between client and contractor functions in order to promote competition – the purchaser–provider split described in Chapter 1. In summary, there are two broad elements to this:

1 Compulsory competitive tendering is being extended to ensure that local government services, from street cleansing to housing management, are provided by contractors rather than directly by local authorities.
2 Responsibilities and financial resources withdrawn from local government are transferred either to services that have been removed from the comprehensive management structure of local authorities to become self-managed, such as schools, or to non-elected public bodies, such as urban development corporations.

Some commentators have suggested that this amounts to a change in the role of local government rather than a diminution of this role. Instead of providing services directly, local authorities will purchase services from contractors and work in partnership with other organisations in their new 'enabling' role. The Association of Metropolitan Authorities has commented that:

> The strategic challenge now is clearly how best to respond to the opportunities to develop a clear community leadership or community government stance whereby elected members can through local authorities influence, shape, negotiate, and direct a whole range of service planning and delivery for which they may no longer have direct control to ensure that this remains relevant, effective and accountable to the local communities which they represent and serve.
>
> (Association of Metropolitan Authorities, 1993b: 7)

Disagreement about the role, resourcing and powers of local councils means that urban policy has been the site of major changes and tensions. Although local government has lost some key functions, most local authorities still maintain an interest in all aspects of the welfare of their local populations. Indeed, the Local Government Information Unit's (1992) core elements of successful local government are the same as the core elements of successful urban policy:

1 Democratic and accountable to local people.
2 Leading with direct services and through partnerships with other agencies to achieve healthy, safe, attractive and sustainable environments.
3 Opportunities for people to make their voices heard about all public services.
4 Shaping, improving and protecting the environment.
5 Equity and equality in access to resources, services and power.
6 Providing links between local, regional, European and global contexts.
7 High-quality, efficient and effective services.
8 Clarity about values, aims and objectives.

The functions of local authorities depend on whether they are district councils, county councils or regional councils (in Scotland). There are 32 all-purpose borough councils in London, together with the City of London Corporation; 36 all-purpose metropolitan district councils in the other major conurbations in England; 39 county councils and 296 district councils in the remaining areas of England; 8 county councils and 37 district councils in Wales; and 9 regional councils, 53 district councils and 3 island areas in Scotland. There is also a large number of community councils in Wales, Scotland and the non-metropolitan areas of England which are consulted about various local issues. In Northern Ireland there are 26 district councils but these have relatively minor powers, most local administration in the province being carried out by appointed boards.

As noted above, in the non-metropolitan areas of England and in Wales and Scotland the structure of local government is currently under review (see Chapter 11). The present two-tier structure of dividing functions between 'lower-tier' district councils and 'upper-tier' county or regional councils is likely to be replaced in most parts of the country with a single-tier of all-purpose local authorities. The government plans that the new unitary councils should be established during 1996–7.

The metropolitan areas of England are the West Midlands, South Yorkshire, West Yorkshire, Tyne and Wear, Greater Manchester, Merseyside and London. Local government has consisted of only one tier of all-purpose local authorities in these areas since the Local Government Act 1985 abolished the Greater London Council and the six metropolitan counties. These 'unitary' metropolitan district authorities have a comprehensive range of local government functions.

The main areas in which local government has responsibilities, although often with other bodies, are primary, secondary and community education; personal social services for adults and children; the provision of council housing and the improvement of unfit private housing; land-use planning and the control of building and development standards; public health, cleansing and refuse collection; local economic development; maintenance and building of highways and road safety; waste disposal; consumer protection; and funding for arts and recreation. Many of these services are provided directly to the public by local government employees but increasingly they are subject to compulsory competitive tendering, with the council responsible for commissioning the service but not necessarily running it.

Much of local government's work is about regulation. The main examples are the control of standards for new buildings and urban development generally, the inspection and control of pollution and food premises, traffic control and road safety.

The absence of county councils in the big English cities such as London, Birmingham, Leeds and Newcastle means that strategic conurbation-wide functions are carried out through joint working arrangements between several districts. These functions include passenger transport, fire and civil defence, economic development, and museums and arts.

Local authorities are assemblies of elected councillors established by statute to carry out the range of local government functions. They generally delegate the management of particular areas of work to committees, such as a housing committee or a social services committee. Elected local councillors (also called 'members') sit on these committees, overseeing the work of professional officers. In urban areas, all but a few councillors belong to national political parties.

Local government employees include managers, specialist professionals such as educationalists, social workers and housing officers, and a large number of administrators and front-line staff to deliver services to the local community. Staff are employed both to advise councillors and to deliver services. They are usually organised into departments, each headed by a chief officer. A chief executive is normally employed to co-ordinate the activities of the various departments. He or she leads the corporate management activities necessary to run a large multifunctional organisation.

The most important reason for corporate management activities is to manage competition for scarce resources. Local authorities are multi-functional, especially metropolitan district councils. Their different departments and services compete for a share of the total local authority budget. This budget comprises central government grant, local taxes and charges (see Chapter 4). Competition is guided by corporate policies which set priorities in the light of political objectives, central government legislation, the overall financial situation, and demographic, social and economic trends.

Large urban local authorities have central policy units to undertake the co-ordination of resource planning across departments. Their job is also to ensure that policies such as equal opportunities are translated into action by departments, to review the performance of departments, and to represent the local authority as a whole in its relations with the public, the media and central government.

Although many local authority central policy units grew out of research and intelligence units, the central policy unit function has now become a separate activity, with research units now found in only a minority of local authorities (Norris, 1989). However, with the increased emphasis on needs assessment and the evaluation of outcomes, research is of increasing importance to local government and urban policy generally (see Chapter 7).

As well as planning and providing/purchasing services for the public, local authorities also act as ambassador and community advocate. They are frequently involved in lobbying central government about issues ranging from unemployment to the National Health Service, and they are represented on many bodies, including urban development corporations, community health councils and school governing bodies. This outside representation has been reduced by central government. In the last few years local authorities have lost rights to membership of water authorities and district health authorities. More recently, however, proposals that police authorities should no longer have a majority of members who are local councillors but should instead have most of their members appointed by the Home Secretary met with widespread criticism and had to be withdrawn (*AMA News*, April 1994).

CORPORATE POLICY

In 1965, Newcastle City Council took the then radical step of appointing a principal city officer as a 'city manager'. Such an appointment was later recommended by both the Maud Committee and the Royal Commission on Local Government in England (1969). The primary functions of what were to become local authority chief executives were to co-ordinate departments at top level and to focus and act on the strategic corporate issues which faced local councils.

Local authorities employ large numbers of specialised professionals, such as social workers, teachers, housing managers, environmental health officers and town planners, all with specific statutory responsibilities to provide particular services. Many issues, however, do not fit into departmental compartments and need to be tackled through different services working together.

Race relations is a specific example of where local authorities have sought a corporate approach to an issue which cannot be 'departmentalised'. Section 71 of the Race Relations Act 1976 placed a statutory responsibility on local authorities to carry out their functions without discrimination or disadvantaging people because of their race. The effect of this legislation was that local councils became the lead local agencies in race relations policy. A number of authorities developed and implemented strategies to combat racism. Specialist committees and units were established. Much more assertive policies began to be implemented, such as evicting council tenants that persistently harassed black neighbours. Some authorities adopted positive action programmes, such as removing racist textbooks from schools.

Before they were abolished in 1986, the Greater London Council and the Inner London Education Authority pioneered the use in Britain of 'contract compliance' as one means of pursuing positive action against racism. Firms tendering for council contracts had to show that they had an equal opportunities policy with regard to their workforces. Such practices met with hostility from both Conservative Party politicians and sections of the press. In 1988 the government responded with a new Local Government Act which stopped local councils taking into account any factors which were not 'commercial' when awarding contracts.

Attempts by some local councils to improve the representation of black employees among their own workforces by taking positive action have had an impact. The London Borough of Hackney saw such measures contribute to an increase in black and minority ethnic representation in its workforce from 12 per cent in 1980 to 35 per cent in 1988 (Sondhi and Salmon, 1992). A number of urban authorities have introduced targets aimed at achieving a representation of minority ethnic staff in their workforces which reflects the demographic profile of their local population.

Another example of corporate policy is the provision of childcare, a very underdeveloped area of provision in the UK, and one that is ill co-ordinated and fragmented. Good provision is essential to realise equal opportunities in employment, education and training, and to help children disadvantaged by poverty or disability towards a reasonable standard of development. There is strong evidence that pre-school provision can compensate to a significant extent for the effects of social and economic disadvantage on later school achievement. But, as well as providing care, effectiveness depends on provision being well-staffed and educationally orientated (Sylva and David, 1990). Local authorities have a key role in researching needs for childcare and providing information about the provision that is available.

Under the 1989 Children Act, social services departments are responsible for regulating child-care provision. Most of this provision is often by the voluntary sector in playgroups, crèches, nurseries, children's clubs and holiday playschemes. Corporate childcare policies can ensure that this role is an enabling one, working within the new regulatory framework. For example, when a local authority provides financial assistance for acquiring or refurbishing a building for community use, it is sensible to consider at the planning stage the standards required by the Children Act, even if child care is not one of the activities planned for the immediate future. This means that grant-aid officers, community workers and planning control officers should all be in a position to advise on these requirements.

Planning for the provision of public services is increasingly becoming a corporate activity. The Children Act stresses that policy must aim for a wide variety of provision for all children in need, rather than a narrow range of specialised services such as residential or foster care. As far as possible, children with 'special needs', such as those arising from disabilities, should not be concentrated in separate, specialised settings, even if this might be the most convenient arrangement for professionals or administrators. A network of child services has to be developed. These have to be run in ways that fit into the broader service provision to families of housing departments, education departments, schools, leisure departments, voluntary organisations and health services.

Increasing numbers of local authorities are reorganising their work to be issue- or people-based rather than service-based. North Tyneside Council, for example, has reorganized its service committees into five standing committees responsible for 'functions'. These are environment, performance review and monitoring, education, social affairs and external affairs. Corporate policy is established by a policy and resources committee which includes the controlling Labour group executive. These executive members meet regularly with the council's executive directorate, five executive directors responsible for corporate implementation of policy, the delivery of services around groups of functions, specific spheres of policy and integrated responses to area-based issues. The policy and resources committee is advised by committees for women, older people, young people, health issues and voluntary organisations.

Kirklees Metropolitan Council adopted a business planning approach during the 1980s. As in North Tyneside, a small group of strategic managers called executive directors ensure that members' policy priorities are implemented across all services. This executive board is essentially concerned with, 'setting the agenda, allocating resources and monitoring results' (Hughes, 1990: 24). Operational responsibility is delegated to 'heads of service' who each have responsibility for a group of services. These organisational changes are complemented by a programme of training and development targeted on frontline staff and managers, and a large community development service. Both Kirklees and North Tyneside have also strengthened political decision-making by concentrating executive responsibilities and power in the hands of a smaller number of councillors. In Kirklees, however, backbench councillors have an investigative role on 'scrutiny commissions' set up by the council to examine issues that have ranged from tourism to low pay.

A number of authorities have adopted a decentralised geographical approach. The most radical example has been the London Borough of Tower Hamlets, which in 1986 disbanded most of the authority's central committees and service departments and passed political and administrative control to seven neighbourhood committees with their own budgets for services. The experiment ended in May 1994 when the Liberal Democrats lost the council to the Labour Party in the local government elections.

Overall, many local authorities now have corporate policies which seek to realise common aims through different services. These are established as *strategic objectives*. The following example from Glasgow City Council illustrates how broadly many local authorities interpret their strategic role:

1 Investing in jobs and sustaining a balanced local economy 'able to compete successfully with other European cities and provide employment for all who require it'.
2 Tackling deprivation and pursuing an anti-poverty strategy by identifying the needs of disadvantaged people and giving these needs 'priority attention across the range of service policies and programmes'.
3 Improving the environment and greening the city.
4 Attracting and retaining population, and securing quality of life and equality of opportunity in a successful metropolitan city, 'a place in which people will want to live and work, and a place which will attract visitors and investors'.
5 Promoting health 'in the broadest sense'.
6 Effectively managing the organisation to achieve quality in directly provided services and through collaboration with other agencies and groups.

Plans for the City Council's individual services are guided by these strategic objectives. Thus, the housing plan is formulated so that it contributes to the overall corporate strategy of the Council. Each of the above corporate objectives is reflected in the following housing objectives: promoting housing investment for jobs; targeting need; improving environmental quality; providing housing opportunities within the city; improving health by improving physical conditions; and working with performance indicators and in partnership with other agencies (Glasgow City Council, 1992). These establish the strategic framework for the housing plan. Its detail is then based on an assessment of needs which considers:

1. Housing needs and conditions in the city.
2. The service developments required to meet the needs of the Council's own housing 'customers' (its tenants and housing applicants).
3. An assessment of requirements for social housing 'client group by client group'.
4. Clarification of the roles of owner-occupation, housing associations, Scottish Homes, the private rented sector and the Glasgow Development Agency.
5. A review of the resources likely to be available.

Following this analysis, a 'realistic and feasible' housing scenario is established. The plan is then implemented through housing policies and programmes for housing management, homelessness, special needs and community care, customer care and investment in renewal. The plan also explains arrangements for implementation and monitoring, and identifies the key issues to be faced, including areas where the Council believes central government policy should be changed.

Three broad themes can be identified in this type of corporate planning:

1. A corporate approach which seeks integration of services on the ground and strategic cross-service planning at the centre of the authority.
2. A community approach which involves consultation and community development, and a wide conception of the local authority's role extending across areas such as economic development and health as well as the more traditional local authority services.
3. Evaluation of outcomes, together with staff development to improve quality, especially from the service consumer's point of view.

Some recent legislative change has required local authorities to establish corporate policies across agencies. The community care changes introduced in April 1993 is a prime example. The key players are social services departments, district health authorities, hospitals, community health services, family health services authorities, and GPs and primary health-care teams. Policy planning must also involve private sector providers of social and nursing care, the voluntary sector, housing departments, and groups representing users and carers. The aim of establishing corporate policies is to facilitate the transfer of long-stay patients from psychiatric hospitals into community-based provision, the quicker discharge of patients from acute hospital care, and the meeting of dual needs for health and social care that many people living in the community have. These activities are all carried out on the basis of a purchaser–provider split between health authorities and health care providers, and between social services departments and social care organisations. They necessitate joint agreements on issues such as divisions of responsibility, resourcing and joint funding, project plans to manage change, and constant networking between officers across agency boundaries.

ADAPTING TO COMPETITION

Community care involves an 'enabling' approach not just because of the co-ordination that is necessary but also because local authorities have to manage a 'mixed economy' of organisations providing care so as to achieve the best possible match with a range of needs for home care, residential care, special equipment and adaptations to people's homes, and community-based services. 'Enabling' has become established practice for housing and planning departments, where public subsidy is used as an incentive to attract private investment into developing vacant land or renewing run-down housing. This is increasingly advocated as a model for all public services. It places local authorities in the role of assessing needs, working in partnership with other agencies, making plans for the provision of services which are not necessarily directly managed by the local authority, and reviewing performance.

The enabling model has much to commend it. But it is widely seen as a smoke screen for privatisation. Government legislation has made it very difficult for local councils' own direct service organisations to win contracts. They are required to achieve a 5 per cent return on capital, whilst private firms can make loss leader bids for contracts and lose money for a time in the hope that they are winning market share for the long term when they can raise their prices.

Another problem is that partnerships between public sector agencies raise problems of accountability (Stoker and Young, 1993). This type of work is often not in public view and responsibility may be ill-defined.

The loss of direct service provision could see the role of local government narrow down considerably to specifying performance targets and awarding contracts. In reality, the extent to which this happens is still likely to depend in part on the policy of the local council. Compulsory competitive tendering is required for road repairs, housing repairs, ground maintenance, cleaning, refuse collection, school meals, the management of sports and leisure facilities, the inspection of schools, careers advisory services, housing management services, computing services, legal services and a number of administrative functions. Many Labour councils have sought to preserve their in-house services in these areas by creating business-oriented direct service organisations managed separately from client departments, and most large contracts have in fact been won in-house. Other councils have embraced privatisation, setting up new units responsible for contract commissioning and quality standards, but withdrawing from direct service provision. Berkshire County Council, for example, recently sold its successful in-house school catering service to a hand-picked multinational catering firm as part of its strategy of divesting itself of running services so that managers can concentrate on objectives, quality outcomes and longer-term strategy. Its chief executive argues that the school catering service is too constrained in the public sector: it had been 'starved' of capital investment, could not access private sector capital, and could not compete for contracts outside the public sector – legislative constraints that had the service in a 'stranglehold' (Allen, 1991).

It is not only direct service provision by local authorities that was reduced during the 1980s, but also their planning and regulatory functions. For example, in many inner-city areas the government appointed urban development corporations (UDCs) to take over land-use planning and development functions from local government. The first UDCs were created in 1981 to tackle what the government claimed were local authorities that were too bureaucratic and had failed to tackle the redevelopment of disused inner-city land. To date, the achievements of UDCs appear to be very modest compared with the size of their budgets (Robinson, Lawrence and Shaw, 1993; Tyne and Wear Research and Intelligence Unit, 1991; see also Chapter 3).

Reductions in the planning and regulatory roles of local authorities continue in the 1990s. Central government has been pursuing a deregulation initiative across all areas of public policy. The Deregulation and Contracting Out Bill, which had its second reading in February 1994, establishes government powers to remove or simplify regulations such as bus licensing, building controls and health and safety requirements. It also enables contractors to carry out statutory functions previously undertaken by civil servants or local government officers. It gives government ministers the power to make orders removing regulations on the private sector which are considered 'burdensome' or 'outdated', provided necessary protection is not sacrificed and key interests are consulted. Instead of the lengthy parliamentary procedures required by primary legislation, draft orders require only a 40-day period to allow scrutiny before they are debated and approved. There has been considerable controversy about the extent of these powers in such a sensitive area involving the health, safety and welfare of the public (*AMA Digest*, 14 February 1994).

The most extensive reductions in the planning and regulatory powers of local authorities have occurred in the area of education. The 1988 Education Reform Act introduced for the first time a National Curriculum prescribed for all state schools by the Department for Education. It strengthened the powers of school governing bodies *vis-à-vis* their local education authorities, and provided for schools to become 'grant maintained' and leave the control of their local education authority altogether. For schools remaining with their LEA, budgets have been devolved to individual schools which now purchase services such as advisers, music teaching and financial services from either the local authority or from another source, including the private sector. Schools are also responsible for buying cleaning, school catering, repairs and maintenance services.

The 1993 Education Act established a Funding Agency for Schools for England and a Schools Funding Council for Wales. Once 10 per cent of school places in a local authority are in grant-maintained schools, these government-appointed bodies become jointly responsible with the local education authority for planning school places. Once 75 per cent of places are in grant-maintained schools the local authority loses all responsibility for planning schools. Given that education accounts for almost half of local authority revenue expenditure, these changes threaten a major reduction in the status and powers of local authorities. To date, however, the policy has had very limited success nationally, with schools preferring to stay with their local education authorities. By January 1994, fewer than 1,000 of the 25,000 state schools in England and Wales had voted to become grant maintained (*The Guardian*, 29 January 1994).

Polytechnics and higher education colleges were removed from the control of local councils in April 1989, taking about £1 billion out of local government expenditure. In April 1993, further education colleges were also removed from local authority control, reducing local council budgets nationally by a further £2 billion. The statutory duty to provide a careers service is being transferred from local education authorities to the Department of Employment, with contracts to provide local careers services being put out to competitive tender.

A large number of local authority services now have to be funded through contracts which are increasingly tendered competitively. In addition, a growing amount of central government funding for local authorities is being allocated competitively. The Audit Commission (1988) has termed the successful council of the future 'the competitive council', for which eight key factors are likely to be crucial. These are worth considering briefly one by one because they map the features of 'post-bureaucratic' urban management, whether performed by local government or other public agencies.

1 Understand its customers

The term 'customers' has been introduced into the public sector in recent years. This development is considered in more detail in Chapter 5. It implies that the only value of public services is the extent to which they satisfy consumer needs.

In the past, estimating needs was a question of forecasting how many houses, beds in residential homes for the elderly, or school places were required. Today, users of public services are more likely to want to be considered as individuals with changing needs, and to be critical about the quality of the services they receive. Local authorities need to understand their 'customers'. The Audit Commission identifies two questions for the local council that seeks to do this:

- Do staff think of the public as customers with views and choices?
- Does the council test customer views and build the responses into policy planning and implementation?

2 Respond to the electorate

The turnout in British local government elections in unusually low; only a little over half of other European countries. This has led to concern about the functioning of local government, and whether links between local councils and the community are strong enough.

These links can be strengthened through measures such as providing more information to the public and improving public participation in decisions. Low local electoral turnouts are probably also a response by the public to the weak constitutional position of local government in Britain and its large size by European standards, making the town hall more distant and with fewer councillors representing larger numbers of constituents (Batley and Stoker, 1991). It has also been suggested that having a system of an elected mayor with strong executive powers would improve accountability and strengthen the internal leadership and management of local authorities (Stoker and Young, 1993).

3 Set and pursue consistent, achievable objectives

Setting clear objectives is a characteristic of good management. It is essential that objectives are realistic and achievable. Objectives must also be consistent with each other.

The growth of contracts between purchasers and providers of public services has made objective-setting particularly important. This has necessitated specifying clear service objectives for contractors to meet which are realistic in terms of the level of resources available.

Objective-setting is also essential in order to bring about change. For example, the London Borough of Brent set the following objectives to ensure that equal opportunities become a permanent part of the local authority's structure:

- Facilitate the recruitment and retention of women.
- Improve the position and working conditions of women.
- Provide career opportunities for women at all levels.
- Increase the representativeness of women in senior positions.

A package of provisions was introduced to meet these objectives (Itzin, 1992). These included a career break scheme, a term-time working scheme, a workplace nursery, and extensions to after school and holiday playschemes. Equal opportunities goals were determined and action plans established, including setting targets for women in senior positions, senior secondment and accelerated management development schemes, together with measures to increase opportunities for women manual workers.

4 Assign clear management responsibilities

Another characteristic of good management is ensuring that departments, individuals and contractors know and agree what is expected of them. Clear management responsibilities have to be assigned, with agreed goals and accountability.

5 Train and motivate people

A key determinant of the quality of any service is the quality of the people delivering it. This depends on good training and high motivation, although a range of factors is involved. These include clear responsibilities, a good working environment, good staff development and appraisal arrangements, and a positive underlying 'culture' for the organisation.

6 Communicate effectively

Assigning responsibilities for what needs to be done, and motivating people about the importance of their work, depends on good communication. This needs to be planned, carefully carried out, and the results monitored. Particularly important for managers is to listen to front-line staff because of the extent of their contact with service users. Good communication is also an essential element in achieving responsive and user-friendly services.

7 Monitor results

The Citizen's Charter has greatly increased the importance for local councils of monitoring key aspects of services (see Chapter 5). Although it can be difficult to monitor some aspects, it is important that councils have means of assessing success or failure service by service, and that action follows from these assessments.

8 Adapt to change

Local authorities have to adapt quickly and effectively to change. They need to invest in monitoring and anticipating change, from the effects of demographic trends to the implications of new national and European legislation and regulations.

Devolution of responsibility has been central to all these changes. Four key elements are emerging:

- Local government services are increasingly being delivered through a number of clearly identifiable business units with devolved management and accounting systems. This can include these units having separate bank accounts.
- Front-line managers and staff are being given clear responsibility for delivering their services, with maximum possible freedom of operation in using resources.
- Separate business units are being given freedom to operate within a limited set of basic corporate rules.
- The traditional function of central staff, such as the internal audit service in making sure that regulations are followed, is being supplemented by a more balanced approach. This includes providing expert advice and servicing for devolved business units.

NEW PARADIGMS FOR PUBLIC SERVICES

One underlying theme in all these aspects of the 'post-bureaucratic' public sector is a retreat from treating whole social groups as having homogeneous needs defined by professional experts. During the 1960s and 1970s, a number of academic studies were made of how urban local authorities were dominated by professional ideologies which treated people as 'policy objects' (Davies, 1972; Dennis, 1972). They argued that policy was based on often vague ideas about the 'nature of the good society' without consulting the people affected. An infamous example of this is the justification of large-scale clearances of working-class housing in the 1970s by Wilfred Burns, chief planning officer of Newcastle City Council:

> The dwellers of a slum are almost a separate race of people, with different values, aspirations and ways of living . . . Most people who live in slums have no views on their environment at all . . . when we are dealing with people who have no initiative or civic pride, the task, surely, is to break up such groupings even though the people seem to be satisfied with their miserable environment and seem to enjoy an extrovert social life in their own locality.
>
> (Quoted in Ward, 1984: 6)

People displaced from such housing areas by demolition were often rehoused in tower blocks and large mass-housing estates built by contractors for the local council. Many communities were broken up and much urban heritage was lost through clearance action, the construction of urban motorways and town centre redevelopment schemes. Despite marked declines in overcrowding and substantial improvements in amenities such as inside toilets and central heating, many people felt alienated by the experience. The following local account is typical of the criticisms made of the mass housing built during this era:

> The council have rehoused people, but that's all they've done ... there are hardly any shops on them even. They expect some sort of respect just because the council build houses. But people weren't allowed to be involved in the creation of the estates, or in the type of housing. People have no say.
>
> (Ayre, 1979: 14)

Economies of scale dominated 1960s planning. In some parts of the country deliberate attempts were made to channel people into large towns and cities where public services could be provided efficiently and where large factories could find ready supplies of labour (Blackman, 1987). The tower blocks and maisonettes of 1960s and 1970s council housing were the standardized products of large construction companies (Dunleavy, 1981). Public services generally became organised into large bureaucracies of professional and semi-professional people both planning and delivering services to their 'clients'. The economic and efficient processing of departmental work and the rationing of scarce resources such as council housing created an often inflexible line management hierarchy dominated by professional officers.

This traditional producer-dominated approach to welfare is not surviving the rise of consumerism and individualism which began in the 1970s. Titterton (1992: 2) criticises the bureaucratic model of working, 'with crude aggregate notions of "problem groups", whose members apparently all share the same fate ... whole social groups such as adolescents and the elderly are by and large still perceived as problem categories in themselves'. Applying this individualist model to social care, Titterton argues that a person's need for social care depends on individual traits and circumstances: how they cope with stress and life events, and the quality and quantity of caring social relationships. Although structural determinants such as deprivation cause threats to a person's welfare, whether these make that person vulnerable to risks such as depression or offending depends on the person's material, social and personal 'coping resources' and their 'coping styles' or methods of coping.

Titterton reflects the contemporary emphasis on the individual in public policy, arguing for an approach to providing welfare services which recognises and caters for differences between individuals, rather than fits individuals into categories based on a few conventional services. Figure 2.1 presents Titterton's analysis in a simple model.

Arguments such as those developed by Titterton reflect the wider shift in public services towards being led by the needs of the individual consumer

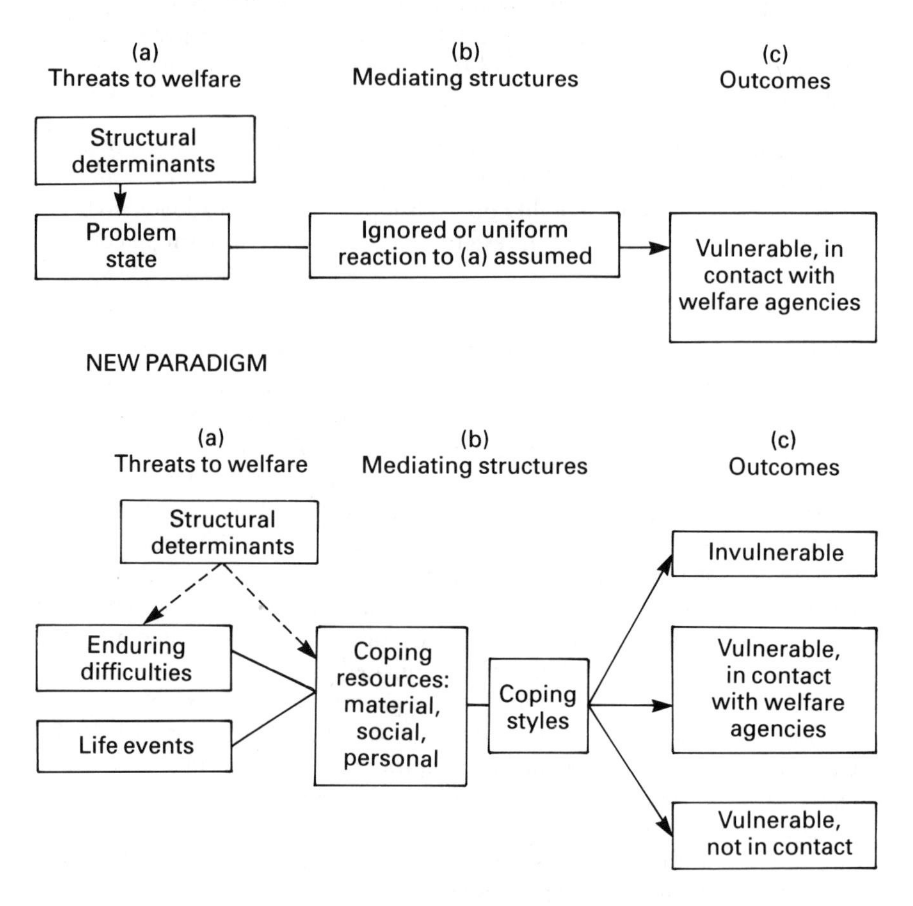

Figure 2.1 Contrasting models of the threat–response process in the old and new paradigms of welfare
Source: Titterton, 1992.

rather than by producer interests. They are very much in line with current practices based on assessments of individuals. These include assessing people's priority for council housing, their eligibility for home care or residential care, the 'statementing' of a child with special educational needs, and means-testing for benefits, grants and charges for services. However, as this list indicates, assessment is as much about gate-keeping access to scarce resources as tailoring social provision to individual needs.

Radical disability organisations in particular have been critical of individual assessment because it perpetuates dependency on social workers, health professionals and caring services (Brotchie and Hills, 1991). Rather than being 'clients' with special needs, many disabled people are arguing that it is society which is disabling them. Buildings and public transport systems are not designed to be accessible to people with disabilities, and there is no equivalent to the Americans with Disabilities Act 1990 in the United States which prohibits discrimination on the basis of disability in employment, public services, transport and housing. Increasingly, disabled people are participating – not just in their individual assessments but in organisations campaigning for legal rights to the same opportunities as people who are 'able bodied'.

Individual assessment for services sits uneasily with local government's traditional concerns with collective needs and aspirations. Whilst some aspects of the emphasis on the individual consumer might be welcome, many people's interests are best served by collective organisation. The involvement of organised tenants, black people, women or disabled people in decision-

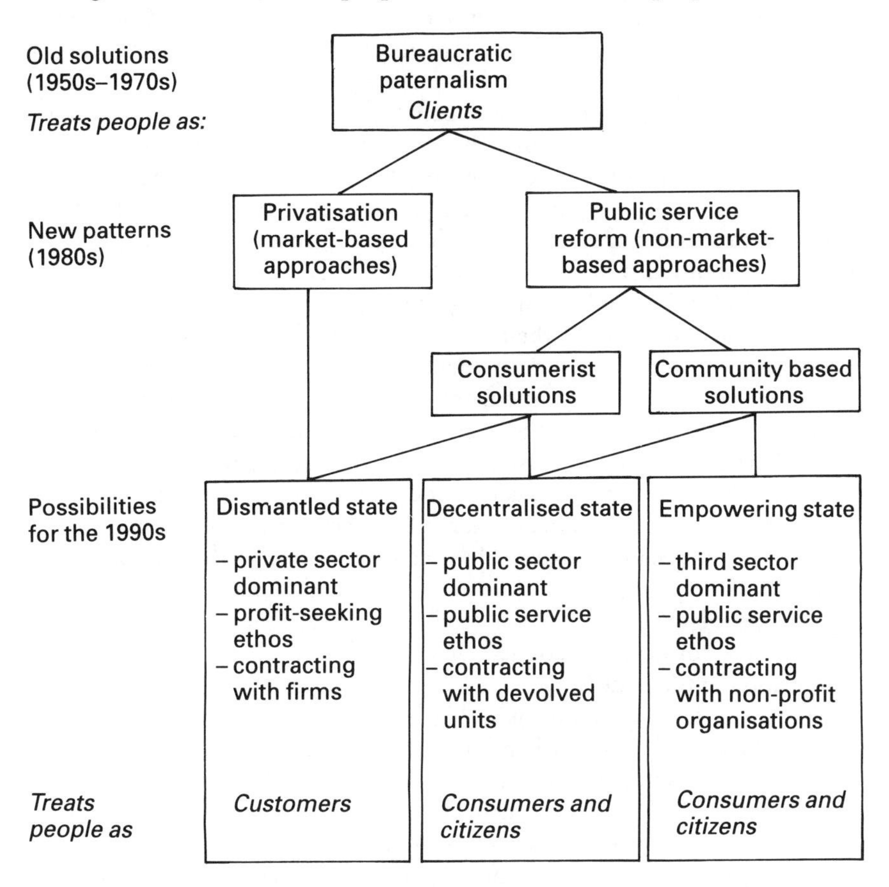

Figure 2.2 A projection of trends in local authority management
Source: Hambleton, 1992.

making is an objective of community development in a number of local authorities, where best practice is to create opportunities for people to control their environments and services.

Hambleton (1992) describes community development as one of three alternatives which have emerged from the rejection of old-style bureaucratic paternalism in local government. The other two are consumerism and privatisation (see Figure 2.2). Privatisation has seen the development of explicit service standards and some moves towards 'customer care', but in a competitive tendering situation it is largely about achieving best value for money. Consumerism is more clearly about 'getting closer to the user', and in public services has been pursued through decentralisation to neighbourhood offices, consumer surveys and user participation in advisory groups. Community development takes this further, possibly leading to the transfer of responsibility for services to 'third sector' voluntary and community organisations, supported and monitored by the local authority. Consumerism and community development are described in detail in Chapters 5 and 6.

Whether through consumerism or community development, best practice in public services now requires that local people are involved as far as possible in identifying their own needs. People increasingly expect to be consulted. This may require the use of advocates for people who have difficulty expressing their needs, such as young children or people with learning disabilities. It involves users of services having representation in the process of planning and monitoring their services. Written specifications need to inform users about their entitlements and how to complain if they are dissatisfied. But even the most open and well-resourced public participation exercises run the risk of frustrating the public when they find that services are constrained by resource shortages and regulations, and often cannot have the flexibility necessary to give consumers 'choice'.

Indeed, the decline of bureaucratic local government has been caused by more fundamental reasons than a search for user-friendliness. Painter (1991) argues that changes in the nature of the economy explain why the role of the local state has changed. By drawing on regulation theory, he offers an explanation of changes that run more deeply than new concepts of the enabling or competitive council. This section concludes with a brief examination of these ideas.

The rise of local government as big business coincided with the period from the 1930s to the 1970s when mass production in large-scale manufacturing plants established itself as the dominant 'regime of accumulation' in the British and world economy. Such plants were typical of the production methods adopted by the most competitive modernising companies at the leading edge of the economy. This regime is often referred to as Fordism, after the large routine assembly lines producing standardised products for mass markets developed by the Ford Motor Company.

Mass production and mass consumption seemed to offer the basis for economic growth on a relatively stable basis after the economic crises of the inter-war years. The state's role was crucial to 'regulating' the conditions of production and consumption that could sustain Fordism, particularly local government as a mass provider of modern housing, social services, public health, education, transport and physical infrastructure. Painter explains this as follows:

> During the period 1950–1970 in Britain, local government came to play a key role in the characteristic mode of regulation of Fordism. There were three areas of particular importance. Firstly, local government provided a range of services whose production was unprofitable for capital under existing technical and organizational conditions, but for which there was a political demand. This principle of universal provision of a range of basic services is one hallmark of the Fordist era ... Secondly, local government was involved in widespread planning and regulatory activity. This included land-use and resource planning and environmental health regulation. Finally, the representative functions of local government meant that the relatively consensual social democratic politics characteristic of Fordism could find partial expression at the local level.
>
> (Painter, 1991: 23)

The regulation theory which guides Painter's analysis is a materialist theory which explains social and institutional change in terms of the organisation of the economy. Thus, the public sector will be expected to reflect the private sector in its organisation and management systems. The evidence seems to support this: the large-scale bureaucracies of local government in the 1950s–60s mirrored the large factories of Fordism. Social services departments were even described as 'Seebohm factories' after the government-appointed committee which reviewed the organisation and responsibilities of local authority personal social services in the late 1960s (Means, 1992).

From the beginning of the 1970s, theorists of Fordism regard the Fordist regime to have entered a period of crisis resulting in a transformation to 'post-Fordism'. Profits from mass production began to decline and the growth of public services became a drain on the private sector. This manifested itself in pressures to reduce public expenditure from the mid-1970s. Intensified competition between private companies led to a series of changes to maintain or increase market share and profits. More highly differentiated products began to be produced with short lifespans, and large companies decentralised production, often to small contractors. New information technology enabled companies to maintain control in a decentralised system. Consumers began to enjoy and came to expect consistent quality and greater choice.

With the election of a Conservative government in 1979, the public sector began to be rolled back to provide more opportunities for private firms to provide public services where there had previously been monopolies for local government employees. Unionised public sector workers were forced to compete with a weakly unionised private sector where wages and conditions were often not as good. Public sector trade unions had achieved for their members wages and conditions arguably superior to what would have been possible had market disciplines prevailed in sectors such as construction and catering where low pay, casual employment and poor conditions are common. The unions had also wielded an ability to have a major impact on the public through strike action because of their monopoly over public services.

In the private sector, the pressures of faltering productivity and loss of markets led many firms to adopt 'post-Fordist' methods such as the more flexible use of labour. Pressure on public spending led to a similar drive to cut labour costs in the public sector through compulsory competitive tendering. Many public services employees have experienced wage cuts, 'speed-ups' and increasing use of part-time and short-term contracts. Thousands of council staff have been made redundant or had to accept worse employment conditions following contracts being won by private companies (although following two European Court of Justice rulings in June 1994, it has been established that the UK breached EU regulations which require staff to be transferred to new employers with existing wages and conditions intact. Compensation claims by employees who as a result received inadequate protection under British law could total £50,000–60,000 (*Financial Times*, 9 June 1994)). The performance of managers and workers has become subject to much more rigorous scrutiny.

Regulation theory is an essentially structuralist theory. It undoubtedly over-simplifies the roles of political ideologies, cultural change and social action, and underplays the achievements of the post-war expansion of the welfare state in terms of raising general living standards and creating greater equality of opportunity. Its value, however, is in recognising how dominant the economic system is in shaping society.

CONCLUSION

From the middle of the 1950s to the middle of the 1970s there was broad agreement about the proper role of local government. The Audit Commission (1988) has summarised this consensus as follows:

1 The public's needs for services such as education, community care and housing were far from satisfied. Services were expected to expand as a result, and the main focus was on numbers: new houses built, college places provided, social services facilities or swimming pools opened. The main debate was about how quickly growth could take place.

2 The best way to provide these services was through the public sector, financed by taxes. Local services were seen as the natural responsibility of local government. The planning and co-ordination of these services was seen as best done by professionals, and this led to a preference for large authorities.

This consensus existed for the first three post-war decades. It began to break down in the late 1970s, and then was directly challenged by the Thatcher government elected in 1979.

Today, the long period of Conservative governments has seen local government rolled back by the expansion of compulsory competitive tendering for public services, the privatisation of council housing, the growth of a mixed economy of social care provision, the loss of powers in education, dependence on private developers and their capital for urban regeneration, and cash limited budgets under pressure to keep pace with new needs. With the abandonment of the welfare state as a means of improving the general standard of living of the whole population, society has become more unequal. The 1980s saw the middle fifth of households in the UK increase their real income by 23 per cent and the top fifth by 40 per cent (Central Statistical Office, 1993). During the same period, the real income of the bottom fifth of households did not change at all.

An important principle of the welfare state was that public services should help compensate for inequalities in income and wealth. The principle of equal opportunities was particularly important; no one should be denied a good education, housing, social or health care because of low income. Increasingly, however, the public sector has reflected rather than compensated for these inequalities. Council housing, for example, has become a welfare sector for households living in poverty. A large number of people now pay for private health care and private schools, seeking such advantages as early treatment and small class sizes. The principle of every child and young person having an equal chance of a good education is being seriously undermined by growing social inequality and the effects of legislation to increase choice and selection. The recent report of the National Commission on Education warned that present government policy of schools competing for pupils on the basis of parents choosing 'good' schools and leaving others to decline is likely to increase inequalities:

> Although in practice this 'choice' is only an expression of preference, we know that it has increased social segregation in schools and widened differences in their performance. There is a serious danger of a hierarchy of good, adequate and 'sink' schools emerging within the maintained system.
>
> (National Commission on Education, 1993: 180)

Many local authorities are strongly opposed to extending policies of competition and choice to the extent that social inequalities and divisions begin to increase. A number of commentators are warning of an arrival in the UK of the levels of ghettoisation, violence and social problems experienced in US cities. Although there is still a substantial way to go, the trends are in that direction.

3

URBAN POLICY AND CENTRAL GOVERNMENT

This chapter considers in more detail the role of central government in urban policy. It begins by considering 'inner-city policy', the specific measures that central governments have taken to pursue urban regeneration. This is followed by a case study of training and enterprise councils (TECs). The chapter concludes by reviewing the 'new public management' which has been introduced both into central government's own departments and into its many executive agencies created to implement national policy locally.

British inner city policy originated with the Urban Programme, introduced in 1968 to provide central government support to local authorities faced with high levels of social need, urban deprivation and racial tension. Its introduction followed a number of studies carried out in the 1960s which identified concentrations of deprivation, social need, housing problems and educational under-achievement in inner cities. During the 1970s, further research stressed the importance of economic and environmental regeneration alongside social programmes (a series of key Inner Area Studies are summarised in Department of the Environment, 1977). Emphasis was also placed on the role of voluntary organisations and the targeting of resources. This research informed the 1977 white paper, *Policy for the Inner Cities* (HMSO, 1977), which widened the scope of the Urban Programme and attached greater importance to local authorities stimulating investment by the private sector in inner city areas. In England and Wales, 41 urban areas were designated for special assistance.

Developments in Scotland followed a rather different course. The Scottish Development Agency (SDA) was established in 1975 as a lead agency for urban regeneration. It worked with local authorities on major urban regeneration projects, most notably the *Glasgow Eastern Area Renewal* (GEAR) project (Donnison and Middleton, 1987). The SDA was merged with the Training Agency in 1991 to create Scottish Enterprise.

With the election of a Conservative government in 1979, inner city policy shifted away from public expenditure on local authority projects to measures which brought the private sector to the forefront of urban regeneration. Local authorities, which were almost all controlled by the Labour Party in inner urban areas, were side-stepped by the new initiatives and branded by ministers as inefficient, bureaucratic and hostile to private business.

The following summarises some of the more important changes which were introduced:

1 In England, a series of urban development corporations (UDCs) were appointed by the Secretary of State for the Environment to take over powers from inner city local authorities to promote property development in declining urban areas. Similar powers were given to the Scottish and Welsh Development Agencies; in Scotland importance was attached to a joint approach with local authorities and other agencies. Although property development has been the prime concern of UDCs, since the late 1980s business expansion and retraining have received more emphasis.
2 Enterprise zones were declared in designated economically depressed areas. The zones were intended to stimulate economic growth by relaxing land-use planning controls and providing tax reliefs for a period of ten years.
3 Urban Development Grants were introduced in 1982, replaced by Urban Regeneration Grants in 1987 and then City Grants in 1988. City Grants are payable directly to the private sector by the Department of the Environment. They subsidise investment in inner city projects which would not otherwise be profitable.
4 The Estate Action programme was launched. This targets special housing capital allocations from the Department of the Environment to improve 'difficult to let' and 'priority' housing estates. Funding is on the basis of partnerships between local councils and private developers.
5 City action teams (CATs) were introduced in 1985 to co-ordinate assistance from government departments to local projects. In 1986, Task Forces were set up to promote enterprise in small areas over a period of two to five years. These are small teams of secondees from the civil service, the private sector, local government and voluntary bodies.
6 Changes were introduced to adapt education to the needs of industry. Compact schemes were introduced in 1988 in Urban Programme Authority Areas in England and priority areas in Wales. These involve young people in working towards goals, such as improved attendance and higher attainment, agreed between themselves, employers and their schools or colleges. In return, they are guaranteed a job with training, or training leading to a job. The 1988 Education Reform Act established city technology colleges with the intention that they attract sponsorship from industry and provide a competitive stimulus for other schools.

Two further initiatives launched in the early 1990s – training and enterprise councils (TECs) and City Challenge – are discussed in later sections of this chapter.

Whatever had been the problems with local authorities, the measures introduced during the 1980s soon attracted criticism, especially in England. The Audit Commission (1989: 1), established in 1982 to keep the economy and efficiency of local government under review, examined the new inner city initiatives at the end of the decade and found that both local authorities and business regarded inner city policy to be, 'a patchwork quilt of complexity and idiosyncrasy'.

These criticisms related to the situation in England. In Scotland and Wales, the Scottish and Welsh Offices are regional arms of central government which work closely with Scottish Enterprise and the Welsh Development Agency to provide a 'one stop shop' to potential developers and investors. In England, the regional offices of central government were essentially regional agents of their separate Whitehall parent departments and this was seen by the Audit Commission as an important part of the problem. Although city action teams were introduced in 1985 to co-ordinate assistance by departmental regional offices to local urban regeneration projects, they were limited in being able to achieve the wider co-ordinating role which the Audit Commission found lacking.

The criticisms did not go unheeded. The government decided that from April 1994 the regional offices of the Departments of Environment, Transport, Employment, and Trade and Industry would be integrated. Ten regional offices of these four departments would each have a senior regional director responsible for all staff in the offices and for expenditure routed through them. The regional offices would maintain close links with local authorities and businesses in their regions. They would also have close links with departments which do not have regional offices, such as the Home Office and the Department for Education.

A key role for the new regional offices is to prepare annual regeneration statements. These are required to set out the key priorities for regeneration and economic development in the region and are applications for resources from a Single Regeneration Budget (SRB). The SRB brings together into one budget the government's existing expenditure on its twenty separate inner city programmes in England, some £1.4 billion in 1994/5. The programmes involved are shown in Table 3.1. The budget has the aim of providing flexible support for urban regeneration and economic development in ways that meet local needs. In fact, almost £1 billion is accounted for by land, property and development-related regeneration, although this may change over time (Stewart, 1994). The SRB has the following objectives:

1. To enhance the job prospects, education and skills of local people, particularly the young and those at a disadvantage in the labour market;
2. To lever in private sector and other resources, including European funding;

Table 3.1 Central government programmes included in the Single Regeneration Budget from April 1994

Department of the Environment
Estate Action
Housing Action Trusts
City Challenge/Urban Programme
Urban Regeneration Agency
Urban Development Corporations
Inner City Task Forces
City Action Teams
Employment Department
Programme Development Fund
Education Business Partnerships
Teacher Placement Service
Compacts/Inner City Compacts
Business Start-up Scheme
Local Initiative Fund
TEC Challenge
The Home Office
Safer Cities
Section 11 Grants (part)
Ethnic Minority Grant
Ethnic Minority Business Initiative
Department of Trade and Industry
Regional Enterprise Grants (plus English Estates to be subsumed into the Urban Regeneration Agency)
Department for Education
Grants for Education Support and Training (part)

3 To encourage economic development and make local economies more competitive;
4 To improve housing through physical improvements, greater choice and better management and maintenance;
5 To promote initiatives of benefit to ethnic minorities;
6 To tackle crime and improve community safety;
7 To improve the local environment and infrastructure;
8 To enhance the quality of life and health of local people;
9 To involve tenant and community participation.

In its first year, 1994/5, the budget operates along the lines of existing programmes. But from 1995/6 onwards a proportion of the budget will be allocated on the basis of competitive bids from the regions as existing programmes and projects gradually expire. The regional offices will identify bids for economic development and regeneration initiatives from local authorities, TECs or other organisations in their local areas which meet the objectives for the budget. These are expected to involve local partnerships.

Each regional office will then submit their recommendations for expenditure to a new Ministerial Committee for Regeneration. This process replaces the Urban Programme and its 57 designated urban areas. All cities, towns and rural areas will be eligible to be considered for support from the SRB.

The Ministerial Committee for Regeneration consists of 10 cabinet ministers together with the Inner Cities Minister. It has been established to consider national regeneration policies, oversee their co-ordination and advise the Secretary of State for the Environment about the resources that each regional office should receive. The Secretary of State is responsible for implementing regeneration policy and is accountable to parliament for the SRB. The committee also gives guidance on regeneration issues to the regional offices and considers matters raised by 'sponsor ministers'. These sixteen ministers have special responsibility for representing and promoting regeneration in local areas of the country.

The Secretary of State for the Environment is the government minister in charge of the Department of the Environment, the most important central government department for English urban policy. Its responsibilities include housing, planning, inner cities, construction, local government, water policy, pollution control, urban and rural conservation, and sport and recreation. In Wales, administrative responsibility for these affairs is devolved to the Welsh Office under the Secretary of State for Wales. In Scotland, most urban regeneration activities are the responsibility of the Industry Department. Scotland has a separate legal system and its own government departments under the Secretary of State for Scotland. Northern Ireland has its own Department of the Environment which is subject to the direction and control of the Secretary of State for Northern Ireland.

Scotland and Wales have seen a trend towards *administrative devolution*. This has involved transferring powers from London to the Scottish and Welsh Offices. The Welsh Office started as a very small body when it was established in 1964, but today is responsible for over 70 per cent of public expenditure in Wales (Osmond, 1993). It exercises its powers through a large number of appointed bodies. However, administrative devolution in Wales and Scotland has been at the cost of local government. In Wales, it has recently been demonstrated that nearly 100 central government appointed bodies or quangos are responsible for just over a third of Welsh Office spending, only marginally less than the entire revenue expenditure of Welsh local authorities (Heath, 1993).

FROM UDCs TO CITY CHALLENGE

Urban development corporations (UDCs) have been the prime vehicle for tackling inner city regeneration since they were launched in 1981. They are powerful 'quangos', run by boards appointed by central government. The extensive powers of UDCs include land purchase, land and property redevelopment, the granting of planning permissions, and loans and grants.

Section 136 of the 1980 Local Government, Planning and Land Act describes the means by which UDCs may regenerate urban areas as follows:

> bringing land and buildings into effective use, encouraging the development of existing and new industry and commerce, creating an attractive environment and ensuring that housing and social facilities are available to encourage people to live and work in the area.

The first UDCs were established in the London Docklands and in Merseyside. By 1993/4 there were twelve in England and Wales with a total budget of £292 million, accounting for over half of the Department of Environment's expenditure on inner city policy. The budgets of individual UDCs ranged from £4 million for Leeds to £96.4 million for the London Docklands. Their spending is targeted on land and buildings, with the aim of 'levering in' private investment. They have a limited life; all the UDCs are due to be wound up by the end of the 1990s.

The largest UDC is the London Docklands Development Corporation (LDDC). Its programmes enabled an extraordinary Docklands property boom to occur in the area during the 1980s. Market-led and largely unplanned, this development outstripped the capacity of local transport links and forced massive additional spending on infrastructure. The Docklands boom collapsed at the end of the 1980s, resulting in dramatic falls in the value of both commercial and residential property, and an oversupply of new office space. Essentially, the main achievement of the LDDC has been to enable an expansion of the City of London's financial sector into the East End, along with the development of mostly up-market housing. Given that nearby areas had high rates of unemployment, the LDDC attracted considerable criticism for failing to take measures such as providing training which would enable local unemployed people to benefit from new jobs (House of Commons Employment Committee, 1988).

The government's immediate response to this criticism was that the prime purpose of UDCs is property development and physical regeneration, and not directly job creation for unemployed local people. As would be expected for a 1980s urban policy initiative, UDCs have to produce performance indicators. These reflect their narrow focus. The six indicators are: hectares of land brought back into use; infrastructure built; square metres of industrial and commercial floorspace completed; housing units completed; £m private sector investment; and number of jobs created. None of these indicators measures impacts on such key inner city problems as unemployment, crime levels or homelessness.

The criticisms made of the LDDC did bring about a shift to varying extents in the priorities of UDCs, with more emphasis from the late 1980s on social rented housing, training for unemployed local residents and community projects. The needs of the local 'community' are now at least acknowledged by all the UDCs.

One of the strongest commitments to a community approach is that of the Tyne and Wear UDC in North-East England, which has stated that:

> In its work, the Development Corporation has two main objectives – to assist the re-creation of a dynamic and prosperous economy in Tyne and Wear, and to integrate developments with the needs of the local community. The two are inseparable. As we physically renew land to create confidence among investors, we also aim to provide real uplift for the people who live here. Local people must participate in the planning process and projects must be relevant to them by providing, for example, affordable homes, leisure facilities, training or job opportunities.
>
> (Tyne and Wear Urban Development Corporation, 1992: 9)

In 1992, this UDC employed five social development staff and had a community development budget of £1 million. As well as grant-aiding community projects such as youth centres and community transport schemes, it has attempted to involve local people in decisions through innovations such as community monitoring panels. These are forums for local people to be consulted about major development projects. However, community development is a minor activity for all the UDCs. Their overriding purpose is to redevelop old industrial land in the inner cities for new uses, often offices and housing. These investments are not targeted on local unemployed people. Job creation has been quite limited, especially when set against the large amount of public expenditure they have focused on these areas (Robinson, Lawrence and Shaw, 1993).

The 1993 Housing and Urban Development Act built upon the UDC model by establishing a kind of roving development corporation, the Urban Regeneration Agency, later renamed English Partnerships. This is a central government agency with a remit to bring vacant, derelict or contaminated land back into use for housing development, commercial and industrial development, recreation or open space. It is intended to complement local government and other urban policy programmes, including City Challenge, Urban Programme, task forces and city action teams. Inevitably, however, this will be a top-down, property-led development agency. It is taking over City Grant, Derelict Land Grant and the work of English Estates in providing industrial and commercial space, mainly in the 'assisted areas'. It has the powers to compulsorily purchase land. It will be able to designate areas where longer-term strategies to tackle widespread dereliction are needed and be able to control development and provide road links. The aim is to work by agreement with local authorities in identifying sites and reclaiming land, and with the private sector in developing the land. Success is to be measured in terms of the amount of land developed, jobs created and who benefits from new jobs.

The agency is another example of a non-elected appointed body being vested with powers previously held by elected local authorities but bringing with it no additional funding. It is under the control of the Secretary of State for the Environment who appoints its six-person board. As with the UDCs, the government's argument is that such agencies can be flexible and fast-moving, forming a centre of expertise with the single purpose of regenerating urban land.

Whilst UDCs and the Urban Regeneration Agency are based on property redevelopment, the City Challenge initiative launched in May 1991 is 'people targeted'. City Challenge was designed to link economic regeneration to creating employment opportunities for residents in high unemployment areas. Its initiatives could be very direct, such as providing cash subsidies to employers who recruit young people from City Challenge areas. In allocating central government resources for area-based regeneration schemes through a competitive bidding process, City Challenge was a forerunner of the SRB (see above).

A key aim of City Challenge was that local authorities should be encouraged to establish partnerships with private sector companies, voluntary organisations, TECs, city action teams, other statutory agencies and local residents. In rounds one and two, local authorities bid competitively for the funding of action plans on behalf of such partnerships. The plans had to demonstrate how public funding would attract further private sector investment and realise a regeneration strategy for the area. The strategy has to be managed and delivered at arm's length from the local council by a board of management with business and community representation as well as local councillors.

In round one of City Challenge, seventeen local authorities from the 57 urban priority areas in England were invited to submit preliminary proposals. Eleven were selected to prepare detailed action plans, which formed the basis of implementation agreements with central government. Starting from April 1992, these round one 'pacemaker' authorities have five years of central government funding at a level of some £7.5 million per year to regenerate designated areas of economic decline and deprivation. The funding is subject to annual review. In the second round, starting April 1993, all 57 urban priority areas were invited to submit bids and there were 20 winners. In this round, local authorities were also invited to make a case for a task force alongside their City Challenge bids. There was no third round for 1994/95 because City Challenge resources were incorporated into the new Single Regeneration Budget (SRB).

The consultancy firm Victor Hausner & Associates advised the Department of the Environment on City Challenge. Their Guidance Note for local authorities submitting bids describes City Challenge as 'targeted area programmes' which would:

1 Develop disadvantaged areas which have significant development potential for the city and are a major constraint on city-wide development.
2 Provide disadvantaged residents with access to opportunities produced by regeneration through specific measures.
3 Link disadvantaged areas and their residents to a city's mainstream economy.

The areas targeted for regeneration had to have 'development potential', and their development must contribute to the broader regeneration of the city. Different policies and resources are then co-ordinated over five years in a concentrated effort to regenerate the area according to a 'vision' for its future.

In contrast to UDCs, City Challenge initiatives must show how disadvantaged residents will benefit from the programme. As with UDCs, quantitative performance measures are required, but these have to cover the range of changes which it is sought to produce in the area, rather than just data about physical change which the UDCs provide. Strategic objectives, programmes, processes and outputs must be specified (see Table 3.2).

The City Challenge money itself is not new money. The annual expenditure of £230 million a year originally came from several different programmes: Urban Programme, City Grant, Derelict Land Grant, City Action Team money, Estate Action, grants to private housing owners for improvements and the Housing Corporation. City Challenge has appeared to be a windfall for local government, but has to be seen against this 'top slicing' of other budgets and the cuts that urban authorities are being forced to make to keep within central government limits on spending (see Chapter 4). For example, Newcastle City Council will receive £37.5 million of City Challenge money over five years from 1992, mostly capital funding, but over the same period is having to cut about £37 million from its revenue budget and additional amounts from its capital spending.

Until 1993, the Urban Programme was the main vehicle for central government support to specific local government regeneration initiatives. Grant aid at 75 per cent was available for revenue and capital spending, although the latter was increasingly emphasised. Local authorities also had to be cautious about revenue projects because funding was time-limited. When grant aid expired they had to find 100 per cent of the funding themselves or close the project. City Challenge projects face the same problem at the end of their five years of City Challenge funding. Resources to extend their work will depend on securing money from local authorities, the SRB or other sources.

The old system of Urban Programme funding terminated after 1992/3, although approvals from previous years continued. A new, much smaller Urban Partnership Fund of £20 million was introduced for 1993/4 and then subsumed within the SRB from 1994/5. Bids from Urban Programme authorities had to be matched with money from the sale of their council

Table 3.2 An example of a strategic objective, programmes, target outputs and costs in a City Challenge action plan

Strategic objective: improve skill levels and access to employment opportunities

Key problems: Post-16 staying on rate in local schools 22 per cent. 20 per cent ethnic minority population, concentrated in one area, with poor English language skills. Training completion rates poor, and access to city-wide employment and FE opportunities hampered by lack of public transport. Knowledge of skills and labour market five years out of date. Lack of customised training – few of those completing training obtain employment.

Key aims: Increase skill levels relevant to known employment opportunities and access to existing city-wide employment, improve travel to city, co-ordinate training and employment services.

Main outputs: New training centre
Improved courses
Additional training places

Leverage (City Challenge funds: private investment): Direct .... Ultimate ....

	Operational objectives		Programmes/ flagship projects	Target outputs: Total	Target outputs: Flagship projects	Costs over five years: City Challenge funding	Costs over five years: Target total Investment levered in (existing plans/commitments)
1	Increase/improve training opportunities	1.1	Positive action training programme	50 places for two years		£50,000	£150,000 (£50,000)
		1.2	Customised training programme	50 new training places			£200,000

Source: Inner Cities Grants Division, Department of the Environment, 1992.

housing and council-owned land. Forty-six bids were successful in 1993/4. The criteria for successful bids included local needs, value for money, the willingness and ability of the local council to use their capital receipts, and the leverage in of private sector and other funding.

The government's competitive bidding strategy for urban regeneration was extended to a proportion of capital financing with the introduction of Capital Partnership, also in 1993/4. This was funded by top-slicing £200 million from existing allocations for credit approvals and government grants to local authorities and £400 million from grants for housing associations. This former included the £115 million Estate Action borrowing approvals for new local authority housing projects. The sum earmarked specifically for local authorities that have sold all of their housing stock was £30 million. The criteria for successful bids in 1993/4 were the extent to which the local authority also financed the project from its own capital receipts and, in the case of a housing association scheme, the extent to which the local authority contributed to the scheme.

City Challenge and these further developments up to and including the SRB, which now brings all urban policy resources together into the same resource pool, are examples of how the allocation of resources has been linked to requirements that local authorities compete for resources between themselves and with other agencies. This compares with the UDC model of simply substituting a new agency for the local authority. Despite the new initiatives, there is less funding with more strings attached:

> To be able to deliver urban regeneration policies in the last decade of the century it is necessary to address the complex of 'people issues'. Housing, health, community safety, education and above all community involvement are now central to the language of regeneration. The fly in the ointment for Labour local authorities has been that the community involvement has to be in the image of the government view of community and individualistic consumer choice.
>
> (de Groot, 1992: 205)

City Challenge gives the involvement of local people considerably greater priority than UDCs or even many Urban Programme projects. Even so, independent research has cast doubt on how effective City Challenge is as a bottom-up strategy (National Council for Voluntary Organisations, 1993). Its stress on efficiency and quick decision-making is likely to emphasise short-term results rather than achieving changes that can have a long-term effect.

City Challenge has been of greater significance as an indication of the wider direction of urban regeneration policy under the Conservatives than as an initiative in itself. This new direction is characterised by four new principles:

1 'Need' is a necessary but not sufficient condition for cities to receive central government funding for urban regeneration and economic development.
2 Agencies which spend public money must also demonstrate the competence to spend the money in ways central government considers appropriate.
3 Public expenditure on urban regeneration should be targeted on where there is development potential.
4 Bids for public money should be subjected to competition.

These principles are evident across the new Single Regeneration Budget, Housing Investment Programmes, local bids to central government for EU structural funds and new 'place marketing' initiatives, most notably the government's recent City Pride scheme which is involving major British cities in making 'city prospectuses' to compete for inward investment and finance for new infrastructure (Stewart, 1994).

The management arrangements for City Challenge also illustrate principles which are likely to be introduced more widely in the public sector. These are:

1 Strong corporate leadership and a reduced role for committees.
2 Practical strategic objectives.
3 Involvement of the local community and the private sector in decisions and in the management and delivery of projects.
4 Sponsorship of community development.
5 The 'bending' of central government programmes to target areas and groups.
6 Monitoring and evaluation which feeds back to management and projects in the 'here and now' and not as reports several months later.

Overall, the thrust is towards 'entrepreneurial cities' which compete nationally and internationally for investment. Government urban regeneration funding is based on removing obstacles to this competitiveness: dereliction, lack of entrepreneurship and poor housing opportunities are seen as holding back potential for economic growth. This has also been used as a justification for keeping social security benefits very low and removing any legal minimum for wages. However, the 'supply side' of local economies includes the level of skills. In the early 1980s this received a new emphasis as a reason why parts of the country were not competing successfully for investment and jobs.

TRAINING AND ENTERPRISE COUNCILS

In fact, skills training had been a national issue in the 1960s when industrial training boards were introduced to help tackle skills shortages during a very different era of full employment. Legislation in 1973 established a national

Manpower Services Commission (MSC), a tripartite organisation to which both employers and trade unions nominated commissioners to work alongside civil servants. However, rising unemployment came to dominate the MSC's training schemes. The MSC itself met with criticism from the Conservative Party and the private sector for lacking private sector leadership and discipline.

A White Paper in 1988 introduced a fundamental overhaul of the British training system. The MSC was replaced and a national network of locally based training and enterprise councils (TECs) in England and local enterprise companies (LECs) in Scotland was established. In 1993/4, there were 82 TECs in England and Wales and 22 LECs in Scotland. The budgets of TECs ranged from £3.7 million to £48.6 million, with an average of £27 million. These are now important agencies in the local public sector, appointed and funded by the Department of Employment and regulated via one-year contracts.

The role of TECs is to increase the skill level of the local labour force by bringing the activities of employers, education providers and trainers closer together. In addition, a wider role is being developed for TECs as local agents in Department of Employment urban regeneration programmes, supplementing the Department of Employment and Department of Trade and Industry fields in which they are already active.

TECs are meant to be private sector employer-led organisations, although many of their staff transferred from the civil service. Two-thirds of their boards are made up of private sector employers, with the remainder drawn from local authorities, education, trade unions, chambers of commerce, minority ethnic organisations and voluntary organisations. Decisions are made in private boardrooms. In summary, their present functions are:

1 To support training and education programmes to meet local targets for qualifications and job placements, and national targets for 'foundation learning' and 'lifetime learning' (see Chapter 8).
2 To plan and deliver Youth Training programmes for young people (mainly 16- and 17-year-olds) and Training for Work (previously Employment Training) programmes for adults that are suitable to the needs of local markets.
3 To encourage employers to invest more effectively in the skills their businesses need by obtaining training commitments from employers (the 'Investors in People' initiative).
4 To bring education provision closer to the needs of the labour market by supporting 'education business partnerships', 'compacts' between schools and local employers, and teacher placements in industry and commerce. Education business partnerships are formal bodies consisting of senior representatives from education and business which are aimed at improving collaboration between the two sectors. Compacts involve

employers in schools, linking young people's agreements to follow personal action plans to subsequent employment and training opportunities arranged with local employers.

5 To raise awareness at local level of National Vocational Qualifications (NVQs), support their delivery and encourage their adoption by employers. NVQs are central to government vocational training policy. They are qualifications designed by employer-led lead bodies and based on standards of competence required for effective performance in the workplace. They are regularly reviewed to ensure their relevance to emerging as well as current skills needs. TECs also promote 'records of achievement' launched in February 1991. These are nationally recognised documents which individuals can use to record their achievements at school and beyond, and plan their personal and career development.

Government employment training programmes dominate TEC provision, although some TECs have sought to fund a greater diversity of training options through part of their budget known as Local Initiative Funds (LIF). TECs also run government enterprise training programmes and can seek to fund projects by applying for grants from the European Social Fund (ESF) and from the government's Ethnic Minority Grants (EMG) scheme.

TECs vary in their organisational styles, degree of local community links and policy priorities (Emmerich and Peck, 1992). All have corporate plans stating their policies. All have attempted to improve access to advice and guidance, some focusing on unemployed people whilst others have opened up training opportunities to people already in employment. Partnerships between schools and business have been funded. TECs have also commissioned research.

Training credits are funded by TECs. These were first introduced in pilot schemes which started in April 1991. Young people who have left full-time education to join the labour market are given an entitlement to train to approved standards. The credit is used by the young person to purchase approved training which is relevant to their employment and career aspirations. The government is extending the scheme so that by 1995 it will encompass all 16- and 17-year-olds leaving full-time education. Following advice and guidance, individuals are funded to take up places with training providers either in work time or their own time. Some TECs require the trainee to repay a portion of the cost of training on entering employment. There is, however, little evidence that training credits have achieved their aim of increasing young people's motivation to train (Tysome, 1993a).

Equal opportunities have not been very prominent in the work of TECs. There is now considerable evidence that equal opportunities is an issue for training which needs a more positive policy response. Both women and young black people appear to suffer discrimination on YT schemes, and young women are concentrated in a narrow range of traditionally stereotyped female occupational skills such as hairdressing, clerical and secretarial work, and child care, all associated with low pay and a greater prospect of unemployment (Lewis, 1992). There is also evidence that women are not gaining equal access to adult training places and that those who do are also concentrated in traditional 'female' skills courses.

Money for TECs is allocated at a regional level using a formula which is related to the numbers of unemployed people and school-leavers. Within regions, individual TECs are only partly funded in this way. There is also an element which rewards performance, such as the number of NVQs achieved, so that TECs effectively compete with each other for a proportion of their funding. As a result, they have targeted measures which get rewarded in this way. Tysome (1993a) cites the example of big cash bonuses that have been offered by TECs to further education colleges over and above the normal payment per student if ambitious output-related targets are met. Output-related funding for TECs is being used to deliver government targets aimed at increasing the proportion of under-19-year-olds who obtain NVQ level 2 standard of training and at trebling to 40,000 a year the number of young people training to level 3 ('A' level standard craft, technician and supervisory qualifications).

In September 1993, the government published its first TEC league tables (Tysome, 1993b). These grouped TECs into four divisions according to performance against seven criteria based upon Youth Training cost, Employment Training cost, NVQs per hundred YT leavers, positive outcomes per hundred ET leavers (i.e. obtaining a job, further training or a college place), NVQs per hundred ET leavers, the number of young people waiting eight weeks or more for a YT place, and success in obtaining employers who make a commitment to the Investors in People scheme (*Employment Gazette*, October 1993). Wide variations were found, although the league tables were criticised for being very crude measures.

Concern has been expressed that output-related funding encourages TECs to select only the 'most employable' for training places and discourages spending on special needs courses and groups, such as young black people, who suffer discrimination in the labour market (Lewis, 1992). TECs have seen significant cuts to their budgets in recent years. If a TEC is to maximise its income it has to be selective and screen out from training places people with poor literacy, numeracy or English as a second language. In addition, cuts in funding make it harder to pay for childcare allowances, thus excluding many women.

TECs have also had problems delivering their mainstream services. Although government policy is to guarantee a YT place to each young person under 18 who is unemployed, the reality is that there have not been enough YT places. In London, the shortfall was estimated at around 22,000 in 1992. Unemployed people under 18 are not eligible for social security benefits, so without a training place a large number of young people face poverty.

Some TECs, however, have sought to address problems of discrimination and disadvantage in the labour market by using part of their funds for training initiatives specifically targeted on women, black people or people with numeracy and literacy problems. Often these are carried out jointly with local authorities. Good practice in this area includes the following elements:

1 Specific research into the training needs of disadvantaged groups.
2 The development of advice and guidance services for adults.
3 Specific outreach and development posts, particularly in the area of basic skills.
4 Foundation and access courses.
5 Courses for women in non-traditional and higher skills occupations.
6 Childcare initiatives.

Women have less experience of, and access to, post-school education and training than men. Many surveys have revealed a high level of demand among women for post-school opportunities. However, women are confronted by several barriers in accessing the opportunities that exist. These include the lack of childcare provision, the hours and length of courses, the image of educational institutions and the costs of attending courses. Access provision is particularly important for women, including short courses, part-time courses and foundation courses which include advice and guidance, confidence skills and study skills. In April 1994 the Employment Department extended its out-of-school childcare grant so that all 75 TECs in England could participate in the scheme. The target is to create 50,000 new childcare places over the next three years. Grant-aided childcare projects are expected to become financially viable after a year.

TECs have been forced to abandon many of their long-term goals in the face of attempting to meet their obligations to guarantee unemployed people training places. This has affected initiatives such as projects for women returners and to meet specific skill shortages. Overall Employment Department expenditure has been cut back sharply, declining by over £2 billion in real terms between 1987 and 1992. Over the period 1983/4 to 1993/4, unit funding for youth training provision fell by 33 per cent and for adult training provision by 49 per cent. These cuts in expenditure appear to reflect Treasury thinking that TEC training programmes are poor value for money when there are few job opportunities at the end of them (Peck and Emmerich, 1993).

The TECs work within a national framework to respond to local needs. Whilst local authorities have found ways of working with these agencies, they mark a further marginalisation of local government by diminishing the role of local councils in economic development and vocational education and training.

A report by Bennett, Wicks and McCoshan (1994) identified a series of serious problems with TECs, including excessive staff time spent accounting to the Treasury and the Employment Department, cost inefficiencies due to high staff/output ratios and the small size of many TECs, insufficient finance for programmes to fulfil their role of upskilling the UK workforce, and a lack of business-oriented staff. The authors also found conflicts between TECs and other business-led bodies such as chambers of commerce, and recommended that TECs and chambers of commerce should be merged into larger bodies, halving the present number of TECs. LECs in Scotland were found to be more successful, largely due to their arm's-length relationship to the government, giving greater flexibility to respond to local needs. Overall, Bennett (1994) estimates that eliminating TEC cost inefficiencies could save around £255 million out of a total TEC budget of £2.3 billion.

A TEC National Council report recently argued that TECs should widen their role by strengthening the part they play in economic development (*Employment Gazette*, February 1994). The report argued for a 'three point approach':

1 TECs should be centres of excellence in services for new business starts, small and medium-sized firms, and technical, management and entrepreneurial training;
2 They should become partners with key local players, including local authorities, in services covering inward investment, finance, business modernisation, technology transfer, enterprise education links, export promotion and import substitution;
3 They should support other agencies where these have a lead role in infrastructure improvement, investment and residential and business premises, and land use planning.

The future role of TECs is likely to be one of planning training programmes within a wider economic development framework. TECs are likely to become strategic partners in local regeneration plans, particularly in association with local authorities and through bids to the Single Regeneration Budget, ensuring that training plays a central part in these plans. The Kirklees and Calderdale TEC has been an example of how this can work. It made a substantial contribution to Kirklees Council's successful City Challenge bid in 1992 and received full backing from Kirklees Council in its bid to the Department of Trade and Industry for a 'one stop shop' facility in 1993. However, this example also illustrates the tensions involved, as the TEC's strategic plan later brought formal objections from Kirklees Council because it claimed functions that interfered with the Council's education service (McBride, 1993).

THE NEW PUBLIC MANAGEMENT IN CENTRAL GOVERNMENT

As well as the creation of local agencies such as UDCs and TECs, the structure of central government itself has changed significantly in recent years. In 1988, the government launched a radical restructuring of the civil service to convert many service delivery functions of government departments into new executive agencies. This was called the 'Next Steps' programme. It continued changes started by the Financial Management Initiative (FMI) in 1979. This sought to improve efficiency and promote a management culture in the civil service by making middle- and lower-level managers directly responsible for budgets and for meeting financial and other performance targets.

The new agencies are headed by chief executives who are employed on fixed-term contracts. For each agency, the chief executive works within a framework of policy objectives and resources set by the minister of the relevant parent department. The agencies have financial and service targets against which performance, and the performance-related pay of senior managers, is appraised. Management has the freedom to determine how to meet these targets within the overall policy and resources framework. The missions and objectives of the new agencies place particular emphasis on the quality of service delivery.

The government intends that 75 per cent of the civil service will be covered by Next Steps agencies by 1998. In addition, many services in government departments are subject to competitive tendering. Overall, the civil service is being reduced to a smaller core. Even in the area of policy formulation increasing use is being made of external advisers.

These changes introduce a 'new public management' which Rhodes sums up as involving:

> a focus on management, not policy, and on performance appraisal and efficiency; the disaggregation of public bureaucracies into agencies which deal with each other on a user pay basis; the use of quasi-markets and contracting out to foster competition; cost-cutting; and a style of management which emphasises amongst other things, output targets, limited term contracts, monetary incentives and freedom to manage.
>
> (Rhodes, 1991: 1)

The new public management has been introduced with particular vigour in the National Health Service. This is not surprising given that the NHS is a very large public service under direct central government control, with a million employees and costing some £35 billion per annum. The core of the organisational changes made in the NHS is the introduction of health care purchasing by health authorities and fundholding general practitioners from an 'internal market' of hospitals and community health services – the purchaser/provider split. The introduction of purchasing is meant to encourage providers to improve the efficiency and quality of their services. This is because purchasing authorities are meant to achieve the best value for money available within the constraints of their budgets. However, for this incentive to work purchasers have to draw up detailed contracts to ensure that standards are specified so that the provider can be held accountable for performance against these standards.

The need to draw up and monitor contracts greatly increases the costs of administration and information processing borne by purchasing authorities. In a study of these 'quasi-markets' in the NHS, Bartlett (1991: 60) concluded that, 'the incentive effects of quasi-markets will need to be substantial if they are to result in a net improvement in the efficiency of health service delivery without themselves being a cause of unwarranted cost inflation'. Two years later, the retiring chair of an NHS trust was quoted as claiming that this was exactly what the internal market was producing. Speaking of the recent appointments of managers and accountants made by both his trust and health authority purchasers, he stated:

> We have put extra layers of bureaucracy, of management, in and we actually do not know if we are getting any more health care out of this so-called market over and above the costs ... We now have to have special contract managers to negotiate with their new contract managers about contracts. We have to have new accountants to negotiate with their newly-appointed accountants to discuss and agree the figures ...
>
> (*The Guardian*, 20 November 1993: 9)

At this stage it is too early to say whether the contract culture is bringing about an improvement in the NHS or other parts of the public sector, where it has been widely introduced. The NHS reforms in England saw 7,610 extra managers and 10,500 extra administrative staff recruited between 1989 and 1991, and a rise in the paybill for managers of 109 per cent to £381 million over the same period. But the Department of Health attributes much of this increase to senior nurses being reclassified as managers and argues that new management has brought benefits to the NHS (*The Guardian*, 3 September 1993).

There have also been advantages and disadvantages for local councils that have been required by central government legislation to separate the roles of purchasing and providing services. Many of these services must be put out to competitive tender (see Chapter 2). Shaw, Fenwick and Foreman summarise recent research findings from the Department which echo some of the experiences in the NHS:

> In terms of advantages, it was accepted that the split served to clarify responsibilities, facilitate the establishment of standards and effective

> monitoring and strengthen management in services that had, hitherto, been under-managed. The disadvantages were viewed in terms of the increased cost of implementing the split, the potential for further complicating organisational lines of communication and, particularly, the likelihood that the split could lead to 'relationships between client and contractor becoming antagonistic and conflictual'.
>
> (Shaw, Fenwick and Foreman, 1993: 22)

The managerial changes led by central government were a forerunner to the national 'Citizen's Charter' introduced by a White Paper in 1991 (Doern, 1993). The Charter applies to the whole public sector, including local government, Next Steps agencies, central government departments, the NHS, nationalised industries, the privatised and regulated public utilities, the courts, the police and the emergency services (it is considered in detail in Chapter 5). A central aim of the Charter initiative is that public services publish performance targets and information on the standards of performance they achieve.

An example is the Employment Service's Job Seeker's Charter. The Employment Service was part of the Department of Employment before becoming an agency. The Charter commits the agency to offering a personal interview to all its 'clients' when a claim is first made, to assistance with drawing up a 'back to work' plan, and to advice on entitlement to benefits (Employment Service, 1991). Staff are trained in customer care to show politeness and consideration. Job seekers can expect to be seen within ten minutes and have telephone calls answered within 30 seconds. Indicators of performance against these and other targets are published, including the number of people helped back to work. The obligations of job seekers themselves are also stated in the charter, including turning up on time for appointments with employers and informing the service if work is found.

CONCLUSIONS

As earlier chapters have discussed, the delivery of urban services is no longer dominated by local authority departments. Urban services today are the responsibility of a *local public sector* which now has many key players: local authority client departments, local authority direct service organisations, UDCs, passenger transport authorities, police authorities, TECs, housing associations, health authorities, health care providers, and educational institutions. All of these are often working closely with firms and organisations in the private and voluntary sectors, either in partnerships or through contracts.

Despite this fragmentation, financial control and policy-making is highly centralised, with considerable power in the hands of government ministers. Urban policy is largely under the direct control of central government and administered by its regional and local offices.

Many services have been struggling with underfunding; resources for the various inner city programmes are set to fall significantly over the next few years, from £1,017 million in 1992/3 to £806 million in 1995/6. Managers in all parts of the public sector must target their services more precisely and demonstrate levels of performance against financial and service objectives. They have to combine requirements of accountability and public service with expectations that their services are competitive and cost effective.

4

PAYING FOR LOCAL SERVICES

All public sector organisations face financial pressures and it is often claimed that public services in the UK are underfunded. In recent years local authorities have been particularly vociferous in arguing that they have been left with large shortfalls in meeting both inflation and new spending needs.

The National Commission on Education (1993) was critical of the UK's 'tiny increase' of 0.4 per cent in education spending between 1980/1 and 1990/1, measured in volume terms (that is, deflating for rising costs so that what is measured is the volume of goods and services that spending buys). It also showed that although public expenditure on training rose by 4 per cent in real terms over the period 1987/8 to 1992/3, it actually fell by 28 per cent between 1989/90 and 1992/3 despite the unemployment rate almost doubling during this time. The problems of underfunding identified by the Commission could be tackled with a relatively modest 0.7 per cent increase in income tax or a 3.5 per cent increase in corporation tax.

Social rented housing has suffered heavily from public expenditure cuts in recent years. New investment in the seven years to 1992/3 was only 60 per cent of what it was seven years previously and, even more dramatically, 16 per cent of its level seven years before that (*Housing*, November 1993). Proceeds from the sale of council houses have been progressively siphoned off to reduce the Public Sector Borrowing Requirement rather than invested in new building. Yet there is a strong economic case for social housing investment. The net cost of £1 billion of public investment in social housing is only half this amount because of the economic benefits that arise from the 30,000 jobs this creates.

The amount of money that local councils are allowed to borrow for all capital investment programmes fell from £3.259 billion in 1990/1 to £1.999 billion in 1993/4 (*Municipal Journal*, 12–18 November 1993). Councils' capital receipts from the sale of assets, which is another way of funding capital investment, fell by more than 50 per cent over the same period, from £2.143 billion to £1.043 billion.

Underfunding is also an issue for the National Health Service. The UK ranks 22nd out of the 23 advanced industrial countries in the Organisation for Economic Co-operation and Development in terms of the proportion of its gross domestic product spent on health services (*The Guardian*, 25 March 1992).

The greatest iniquity, however, exists for people who have to depend on social security benefits. Under recent Conservative governments the value of benefits has fallen considerably relative to general living standards as a result of being frozen in real terms whilst earnings have increased.

Higher levels of public expenditure mean higher taxes. The need for this could be reduced if the UK improved its inflation and trade performance to boost employment in the market sector of the economy, thus reducing expenditure on unemployment-related benefits and increasing tax revenues. However, although it is widely regarded as fair that the well-off should pay more in taxes, even maintaining the UK's welfare state in its present form will require higher taxation for most of the population, at least until market sector employment expands significantly (Rowthorn, 1993).

Although local authorities and other public bodies participate in national assessments of pressures on spending to finance public services, the planning of public expenditure is a top-down process by central government. government departments have to make a case for spending in competition with each other. The economic climate and the Treasury's assessment of what is 'affordable' are key determining factors in how spending decisions are made.

The funding of any public sector organisation is a complex matter. This chapter focuses on local government to illustrate both key aspects that apply to any part of the local public sector and the particular issues which relate to local authorities.

FINANCIAL PLANNING AND BUDGETING

Financial planning in public sector organisations is driven by two main considerations: *allocating* resources in line with chosen policies and priorities, and *controlling* the use of these resources. In the case of local government, there is the additional need to decide the level of local taxation necessary to pay for public services and capital projects.

These considerations are the main reasons why public sector organisations prepare annual revenue budgets. However, the annual preparation of a revenue budget is itself insufficient as a mechanism for either financial or policy planning. Consequently, many public bodies use *medium-term plans* to provide a context for annual budgets, particularly to set out strategic directions for the organisation. These link the organisation's policy priorities to its revenue resources, usually over a period of three years. This medium-term horizon is required in order to plan for change rather than just make short-term reactions to events.

Changes in the need for public services occur over a relatively long period.

The prime example is demographic change, such as increases or decreases in the number of children needing school places, or in the number of older people needing social care services. Demographic change will also affect the resources available to fund services because central government funding to local authorities, health authorities and other public bodies is largely based on the size of their local populations.

The management of changes in service provision is therefore a medium-term process which responds to new needs and to increases or decreases in resources. Medium-term plans have increasingly been adopted in recent years to avoid the dislocations and unnecessarily high financial costs which can follow from a short-term perspective. They are vehicles for planning the growth, maintenance or decline of spending. They are also used to plan for change introduced by new legislation.

The medium-term plan can serve as a strategic plan for the organisation. Alternatively, it can be a separate document which links the strategic plan to the allocation of resources. Either way, it is a corporate plan for the whole organisation. Within this, each service will often have a business plan, linking its forward development to the overall policies and resources of the whole organisation (see Chapter 5).

The main elements of a local authority medium-term plan are as follows:

1. Forecasts of financial, population and social trends, and the effects of legislation and other changes.
2. An expression of underlying values, objectives and sense of direction. This could include commitments to prioritising resource allocation towards tackling the effects of deprivation, improving the quality of front-line services, or investing in economic regeneration.
3. Consideration of organisational responses to anticipated changes. Certain services, such as social services or economic development, might be prioritised for growth or protection from cuts. New strategies might be set out, such as introducing a split between client and contractor in service provision.
4. An examination of the options, contingencies and priorities for the development of the organisation and its services. This will include options for financial growth and expenditure reductions. There has also been a growing trend for public sector organisations to consider how their services interrelate. This is necessary to reduce duplication and to improve co-ordination, making the most effective use of resources.
5. Establishing overall financial targets and then allocating resources, including planning growth or reductions, to each service area.

The revenue budgets of departments within the organisation are usually decided annually and within the context of the medium-term plan. Departments' detailed estimates will normally be prepared within corporate guidelines from the central finance department, with a process of negotiation before the total revenue budget is decided. Support services such as personnel or computing now generally have their budgets determined through service level agreements with direct service departments, or are provided by external contractors (see Chapter 5).

Newcastle City Council has used medium-term planning to manage decline. It is an urban local authority in an economically depressed region. A continuing loss of population has combined with central government capping of its budget to force reductions in revenue expenditure on services, despite substantial social and economic problems. Census data shows that between 1971 and 1981 Newcastle's population fell by 11 per cent. Between 1981 and 1991 it fell by a further 5 per cent. In 1991, unemployment was 19.3 per cent, and 29.5 per cent of children lived in households where there was no wage earner. Sixteen per cent of residents had a limiting long-term illness or disability.

Medium-term planning is used in Newcastle to manage the reduction in spending required by central government in a way that limits its impact as far as possible on low-income households and deprived areas (Newcastle City Council, 1993a). The overall financial reduction target the authority adopted for the three-year period 1993/4 to 1995/6 was £25 million. This target was then allocated pro rata to departments so that each service's target share of the reductions was in proportion to its total revenue budget. This method was used as the framework for departments to demonstrate where they would make cuts in their services to meet their target, including allowing for any increases in spending. The final decisions on these proposed cuts are taken by the council's majority political party. Political priorities adopted at this stage mean that the final distribution of cuts between services is not pro rata, so that some services are 'reprieved' whilst others have to find larger cuts than their pro-rata share.

One of the political priorities of Newcastle's ruling Labour group on the council is to protect the provision of quality front-line services to high priority groups. As part of the annual budget cycle, the council's anti-poverty advisory sub-committee undertakes a poverty impact analysis of the budget reductions proposed by departments. Table 4.1 shows an extract from the analysis for the 1993/4 medium-term plan. The sub-committee identifies which proposed cuts it advises should not be implemented and this is then taken into account in the final political decisions about budget reductions. Other considerations also impinge on these decisions. Areas of spending which are not required by law are particularly vulnerable to cuts. This was the case with school clothing grants to low-income parents for example (see Table 4.1).

The annual financial cycle in the local public sector begins in the autumn when work starts on preparing revenue budgets for the forthcoming financial year. This is done within given criteria, such as estimates of pay and price inflation ('out-turn' levels of pay and prices), the level of central government

Table 4.1 Newcastle City Council Anti-Poverty Audit: 1993/4 reductions affecting school-age children

Proposal		*Target*	*Comments*	*Decision*
1	Reduce schools budget	£2,150,000	Will affect schools in broadly uniform way	£1,650,000
2	Reduce Education Psychology Service	£100,000	Significant reduction in preventive and case work and in-service training. Will affect special-needs pupils but not statemented pupils	£0
3	Reduce Education Welfare Service by half	£220,000	Significant reductions in support to schools, parents and students. Truancy work will not be lost	£80,000
4	Reduce Education Support Service	£200,000	Reduces major support network used by low income families	£0
5	Reduce Hospital Teaching Service by half	£140,000	Service would concentrate on long-stay patients	£75,000
6	Reduce Instrumental Teaching Service	£200,000	To be covered by selling service to schools	£200,000
7	Foreign Language Assistants	£75,000	Total reduction in service	£75,000
8	Advisory Teachers	£130,000	Schools to buy back services if required	£135,000
9	Education Development Centre	£50,000	Income target to be increased	£50,000
10	Education Support and Training	£100,000		£100,000
11	General Advisers	£110,000	Implications for inspection under 1992 Act	£110,000
10	Discretionary awards	£600,000	Fewer students from low income families will have access to further education and reduce their opportunities to get out of poverty trap. Demand increasing	£0
12	Clothing grants – delete budget	£420,000	Significant effect on families on low incomes. Could increase debt and lead to truancy	£420,000

Source: Newcastle City Council, 1992c.

grants and income from local taxation. Many items of expenditure can be forecast with some accuracy, including for example the costs of employees, but others have to be informed guesses, such as expenditure on housing benefit which will vary according to the number of claimants.

Public sector organisations have increasingly adopted two related measures to avoid overspending. The first is to cash-limit budgets and the second is to adopt cost-centre management. Together these involve designated managers having responsibility for keeping their devolved budgets within given expenditure limits. Pressure on resources has meant that budgets have to be scrutinised so that unnecessary expenditure is minimised. Sophisticated methods of financial planning such as multi-year budgets, planned programme budgeting systems (PPBS) and zero-based budgeting (ZBB) have been adopted by many large organisations in both the public and private sectors. Zero-based budgeting has attracted growing interest. It involves a full critical analysis of expenditure in each area, and may result in radical changes in the distribution of resources between programmes if spending patterns are out of line with policy priorities. This contrasts with more traditional 'incremental' budgeting where, instead of starting from a 'zero base' assumption, financial planning starts from last year's budget and usually makes only small upwards or downwards adjustments.

Spending on public services is largely 'revenue' spending. This is principally the pay of employees but it also includes short-term outlays such as office supplies and fuel. Spending on building and refurbishment projects is classed as 'capital' expenditure. This is long-term investment in what are often very expensive assets such as schools, housing and roads, financed largely from loans repaid at interest from revenue budgets over long periods of time. Capital expenditure is essential to safeguard the efficiency and quality of services. This includes the replacement of vehicles and equipment as well as new building and refurbishment. Failure to make capital investments runs the risk of having to incur higher revenue expenditure on maintenance and repairs in the future.

The capital programmes of public sector organisations are funded in four ways: (a) by government and European grants; (b) by using income from the sale of land and buildings; (c) by borrowing; and (d) directly from revenue accounts. Capital expenditure is controlled by central government. Expenditure by local authorities, for example, is controlled by the government determining how much each local authority is allowed to borrow. It does this by issuing Annual Capital Guidelines to each local authority. These are made up of: (a) permissions to borrow, and (b) the proportion of the local authority's capital receipts which it will be permitted to spend on capital projects. Permissions to borrow consist of Basic Credit Approvals (BCAs) for mainstream services such as housing or education, and Supplementary Credit Approvals (SCAs) for specific items of expenditure such as Estate Action projects or road schemes.

Urban Development Grant, introduced in 1981, used to enable local authorities to subsidise private sector capital projects in economically depressed areas. In 1988, UDG was recast as City Grant to be given directly by central government to businesses, thereby cutting out the local authority role. Derelict Land Grant, introduced in 1982 to bring derelict areas into use, is available to local authorities and to private and voluntary sector organisations.

Managing capital programmes can be difficult because physical progress on schemes such as building projects is prone to disruptions, such as bad weather or shortages of materials. This means that timetabling ('programming') and carrying out regular cash flow forecasts are very important.

LOCAL GOVERNMENT FINANCE

Most of the public sector is funded directly by central government from national taxation, but local authority spending is also financed from local taxes. Central government funding to local authorities is largely Revenue Support Grant (RSG) but also some special grants for specific purposes. Local taxes are the uniform business rate and the council tax. The sources of local authority income in 1993/4 are summarised in Table 4.2.

The uniform business rate is a national tax on commercial and industrial properties. Non-domestic rates used to be levied by local councils on commercial and industrial properties in their areas, and varied from district to district. During the 1980s they attracted considerable criticism as unfair burdens on business, especially in inner cities where councils had increased business rates to compensate for cuts in central government grant. Although this criticism did not bear up to examination, the power to levy non-domestic rates was removed from local authorities with the reforms in local government finance that were introduced in 1990 (Stoker, 1991). However, the transfer of this tax to central government did not go smoothly because it was accompanied by a revaluation of non-domestic properties. This substantially increased the rates bills of many businesses, especially in the south of England, adding to the political damage already caused by the poll tax.

Table 4.2 Sources of local authority income (1993/4)

	£ billion	*Percentage*
Council tax	8.0	19.4
Non-domestic rate income	11.6	28.2
Revenue support grant	17.0	41.3
Specific grants	4.0	9.7
Community care grant	0.6	1.5

Source: Derived from Audit Commission, 1993a.

In 1993/4, non-domestic rates met 28 per cent of council spending. Their removal has thus taken away from local government a substantial degree of financial autonomy. Restoring locally set business rates would increase the proportion of spending controlled by local councils themselves from around 19 per cent to 47 per cent of net council expenditure.

Central government grants account for the highest proportion of local government income. As a proportion of general current income from government grants, non-domestic rates and council taxes, they increased from 42 per cent to 56 per cent between 1990/1 and 1993/4. There was a corresponding decline in the significance of income from council tax and its predecessor, the community charge or 'poll tax'. Local authorities' budgets are thus largely determined by central government assessments of what they need to spend (which are called 'standard spending assessments' or SSAs). Central government also caps local authority budgets if they qualify as 'excessive', limiting the amount of money councils can raise through the council tax. The overall effect is that central government's assessment of what a council needs to spend is now far more important than the council's own estimates of its need to spend. Local councils have become increasingly like other public sector agencies with their budgets determined by central government.

Local councils receive income other than government grants and local taxes. The most important are income from council housing rents and capital receipts from the sale of council houses and other council property.

Local authorities are only permitted to spend on new capital expenditure 25 per cent of their capital receipts from the sale of council houses and 50 per cent of their capital receipts from the sale of other assets such as land. The other 'reserved' portions have to be spent on repaying debt, financing capital expenditure authorised by a government credit approval, or banked if unused. There was an exception to this between November 1992 and December 1993 when the government permitted councils to spend all of their receipts from the sale of council housing and other assets in an effort to stimulate the construction industry. It did not, however, sanction the release of some £5 billion of capital receipts accumulated from before this date.

This limitation on spending capital receipts is opposed by many local authorities which want to reinvest the money in social rented housing. This limit was imposed by government in order to reduce the Public Sector Borrowing Requirement, the government debt. Every pound from the sale of capital assets is counted as 'negative' public expenditure to offset the PSBR, in the same way as other privatisations. In fact, council house sales have been the government's most successful privatisation, raising £28 billion.

STANDARD SPENDING ASSESSMENTS

For more than 150 years central government has distributed grant-aid to local authorities. There are three main reasons why central government funding is necessary. First, it ensures that services of national importance are provided appropriately at local level; second, it supplements local taxes with national revenues and thus reduces the financial burden on local authorities; and third, it gives additional assistance to local authorities with less resources or higher needs, enabling them to provide similar services to better-off authorities without imposing a heavier tax burden.

Until 1958, grants were given mainly for specific services. Since that date, most grant-aid has been distributed in a block, allowing local authorities to decide themselves on the distribution of money between their services.

Many radical changes have been made to local government finance since the 1950s. These are summarised in Table 4.3.

Standard spending assessments (SSAs) were introduced in 1990 as a means by which central government allocates money to local authorities in England and Wales using indicators of the 'need to spend'. They were a simplification of the Grant Related Expenditure Assessments which they replaced. The English and Welsh SSA systems are different although similar principles are involved. The Scottish equivalent is the Grant Aided Expenditure (GAE) system.

SSAs are an important example of *formula funding*. They are used annually

Table 4.3 Government grants to local authorities: major changes since 1958

Year	*Changes*
1958	Many specific grants consolidated into one total grant which councils could choose how to allocate between services.
1967	Rate Support Grant introduced, including the first direct subsidy for domestic ratepayers.
1974	Needs element of Rate Support Grant calculated using regression analysis of actual spending.
1981	Needs and resources elements consolidated into Block Grant. Grant Related Expenditure Assessments (GREAs) introduced.
1985	Local authority spending directly limited by central government through rate capping, initially for selected authorities.
1990	Introduction of the Community Charge (poll tax) and National Non-Domestic Rates. Block and Domestic Rate Relief Grant replaced by Revenue Support Grant, based on Standard Spending Assessments.
1992	Expenditure capping extended to all local authorities.
1993	Council Tax replaces the Community Charge.

Source: Adapted from Audit Commission, 1993a.

to allocate a very large amount of public spending entirely on the basis of statistical indicators. As such, they have the apparent advantage of linking resource allocation to indicators of need, but they have the disadvantage which results from the difficulty of finding valid indicators of need which are measurable *and* available for all local authorities.

Formula-based systems of resource allocation are also used to distribute central government funding to health authorities for hospitals and community health services, to family health service authorities for GPs and dentists, to local authorities and the Housing Corporation for capital funding for social housing, and to training and enterprise councils (TECs). For housing and TECs, only part of the allocations is formula based; part is also allocated on a discretionary basis (see pp. 57, 94–9).

Formula-based resource allocation to health authorities is modified to ensure that allocations do not diverge too far from historic levels of funding, maintaining stability and avoiding disruptive changes in funding levels. However, in local government SSA allocations do change significantly from year to year. Large changes have the effect of forcing authorities to make big upwards or downwards changes to their council tax because of the effects of 'gearing' and government capping of budgets (see pp. 75–6, 89–90).

Understanding how SSAs work is an essential part of understanding the profound effect which central government has on local councils and their services in Britain:

> SSAs may be a remote concept when considering the numbers of pupils per teacher in schools, the extent to which home helps are available to the elderly, the frequency with which potholes in the road are repaired and the availability of new books in the library, to name but a few examples, but all of these services compete for money locally from a budget which is constrained by the SSA system. Directly or indirectly SSAs affect everyone.
>
> (Audit Commission, 1993a: 14)

Councils are allocated their SSAs in December each year for the forthcoming financial year. For 1993/4 central government's assessment of the total need to spend by all local authorities in England was £41.2 billion. For 1994/5 it was £42.7 billion. This amount is known as 'Total Standard Spending' (TSS) and is set out annually in the RSG settlement. TSS takes into account an assessment of the various pressures on local authority services, any changes in local authority functions or new responsibilities, and the potential for improved efficiency in service delivery. The amount the government contributes to TSS from national taxation is the balance between TSS and the level of council tax it considers reasonable for each local authority to levy. This is known as the council tax for standard spending (CTSS). CTSS is broadly the same for each area of the country – in 1993/4 a band 'D' council tax of £494. In 1994/5, CTSS would raise £8.4 billion if councils levied their council tax at this level, so the government set its contribution from national taxation at £34.3 billion (which is called 'aggregate external finance' or AEF). This is in the form of RSG, national non-domestic rates and various special grants.

The 1993/4 settlement was £2.2 billion less than the amount forecast by the local authority associations as the necessary level of spending on local services. The 1994/5 settlement was £3.6 billion less, a shortfall of some 8 per cent compared with the amount local authorities estimated to be their real expenditure needs (Association of Metropolitan Authorities, 1993a).

The RSG settlement is not itself determined by the level of need, although its *distribution* among local authorities does use a needs-based formula. The size of the national cake to be distributed to local authorities depends on political considerations about the level of taxation and borrowing, which are strongly influenced by the Treasury and economic conditions.

RSG is a general grant towards local services which reflects differences in needs *and* resources across local authorities. In order to calculate the amount of grant to be paid to a given local authority, the Department of the Environment first calculates the SSA for that authority. This is the Department's assessment of the budget required to provide a 'standard level of service' consistent with keeping within the national level of public expenditure the government is prepared to incur.

The RSG entitlement for each local authority is the difference between its SSA and its estimated sources of income from council tax and national non-domestic rates. As noted above, the government uses a figure for council tax which it decides is appropriate for the local authority to levy (CTSS) and which is broadly the same for all authorities. It also decides the rate poundage for national non-domestic rates. RSG is fixed at a level where if all local authorities spend in accordance with their SSAs they could each levy broadly the same 'standard' council tax. If a local authority spends higher than its SSA, its council tax will have to be higher than the CTSS, although this is tightly constrained by government capping of local authority budgets (see pp. 89–90).

A local authority's income from the CTSS will vary depending on the value of domestic properties in its area (its tax base). By taking the difference between the SSA and local income from the CTSS, the RSG compensates areas with smaller tax bases. This is known as *resource equalisation*. Local authorities with a larger proportion of domestic properties which are valued relatively low, thus raising a similarly low amount of council tax, receive bigger revenue support grants to compensate.

In summary, the RSG for each local authority is calculated using the following formula:

RSG entitlement = SSA – share of national business rates – (standard tax × council tax base)

If a local council wishes to finance a higher level of service than the SSA, it is constrained by two controls. First, any increase in expenditure above SSA has to be met entirely by the local council taxpayer because RSG is fixed in relation to the SSA, not actual spending. With CTSS meeting about 20 per cent of total standard spending, on average for each 1 per cent variation in spending a 5 per cent variation in council tax results. This is called the *gearing effect*. It is a major disincentive to councils wanting to increase spending above SSA, and has a particularly marked impact on local taxpayers in authorities with high SSAs (see Table 4.4). Second, central government 'caps'

Table 4.4 The gearing effect

	Local authority spending per band D property (£s):			
	Mole Valley	*Doncaster*	*Tower Hamlets*	*England*
SSA	1,392	2,758	4,733	2,247
Council tax for a band D property	493	493	493	493
Extra spending (10%)	139	276	473	225
Council tax paid with 10% extra spending	632	768	966	717
Percentage increase in tax rate	28%	56%	96%	46%
Gearing factor*	2.8	5.6	9.6	4.6

Source: Audit Commission, 1993a.
Note: *Percentage increase in local taxation for each 1 per cent rise in spending.

Table 4.5 Service 'control totals' for England, 1993/4

	£m	*% total SSAs*
Education	16,531	45
All other services	7,007	19
Personal social services	4,902	13
Police	2,788	8
Capital financing	2,520	7
Highway maintenance	1,736	5
Fire and civil defence	1,139	3
Total SSAs	36,623	100
Plus		
Specific grants	3,981	
Community care	565	
Total Standard Spending	41,169	

Source: Audit Commission, 1993a.

council budgets so that councils have very limited freedom to tax above the level that the Secretary of State considers to be reasonable (see pp. 89–90).

The consequence of these two controls is that local council spending is tightly controlled by central government.

SSAs are built up from assessments of the need to spend on major service blocks. Housing is excluded and is subject to a separate allocation process (see pp. 94–9). As Table 4.5 shows, education is by far the largest SSA block, accounting for 40 per cent of Total Standard Spending. Although calculations are made for each of these blocks and sub-blocks, it is the total SSA which is used for distributing RSG. In general, local authorities have discretion over their spending priorities both between and within services.

The calculation of SSAs for each block of services is carried out by first identifying the appropriate *client group*, which as far as possible represents the recipients of the service. For the highway maintenance block this is not appropriate and a measure based on road lengths is used. For other services, the need to spend is assessed by multiplying the relevant client group by a basic unit cost per client and, for certain authorities, an area cost adjustment. These costs are worked out from actual local authority expenditure. The area cost adjustment principally allows for higher labour costs in London and the South-East. In summary:

$$\text{Service SSA} = \text{client group} \times \text{unit cost} \times \text{area cost adjustment}$$

The cost per client is based on the basic unit cost plus variable elements which take into account additional needs. The distribution of these additional needs among local authorities is measured using proxy indicators. These adjust SSAs for additional costs arising from these needs. For example, in the case of primary education the need to spend is mainly based on the local authority's number of pupils aged 5 to 10, a straightforward client group. The cost per pupil consists of a basic annual unit cost plus variable elements. These variable elements increase the basic unit cost to recognise types of additional expenditure which result from the provision of services to meet particular needs.

In both the primary and secondary education elements of the SSA three factors are included in the calculation of additional needs. These are:

1 An *additional educational needs index* which measures the additional cost of meeting the needs of pupils who experience difficulty with learning. Such difficulty can arise for a variety of reasons, including minor or temporary learning difficulties, physical or sensory disabilities, emotional or behavioural disorders, having English as a second language and the effects of social deprivation. About 2 per cent of the school population is thought to have severe and complex problems (National Commission on Education, 1993). These are usually formally assessed and a statement produced describing the child's needs and the provision

which should be made. The 1980s saw a trend towards greater integration of children with special needs into ordinary schools, with less use being made of special schools. Ordinary schools also have to provide learning support resources for a further 18 per cent of children who are thought to experience at some time in their school career difficulty with learning for a variety of reasons. The additional educational needs index adjusts the basic unit cost on the basis of the following indicators of special needs: the proportion of children living in lone-parent families, the proportion of children dependent on Income Support claimants, and the proportion of children with heads of household born outside the UK, Ireland, the USA or the Old Commonwealth. These indicators are used, rather than direct measures such as the number of children in special schools or who are 'statemented', because the latter will vary according to local policy and practice and represent only part of the range of additional educational needs.

2 A *ward sparsity index* which adjusts the SSA for the extra costs of provision associated with the proportion of a local authority's resident population living in wards with low population densities.
3 A *free meals index* which takes into account the requirement that local education authorities provide free school meals to children from families on Income Support.

Weights are given to these different indexes based on judgements about the proportions of spending attributable to them. For instance, the government's opinion is that about 20 per cent of primary and secondary school costs are attributable to the additional educational needs index. The weight given to the index is designed to ensure that about 20 per cent of the primary and secondary sub-blocks of SSAs are distributed on the basis of this index.

An important method which is used to develop this type of indicator is *regression analysis*. This statistical technique helps to identify whether or not relationships exist between sets of data. It is used (a) to identify those social and economic variables which help to explain variations in need to spend, and (b) to define the relative weights that are applied to the variables used.

The need to spend is often measured by actual local authority expenditure, but other measures – such as number of clients – are used in some cases. In the example of the additional educational needs factor described above, regression analysis was used in its development to explore the relationship between possible proxy indicators of the need to spend on educating pupils with special needs, and variables such as teacher–pupil ratios and children with special needs statements which are treated as direct measures of provision for this client group (although it is debatable whether teacher–pupil ratios are an indicator of educational need). For example, in searching for possible proxy indicators, housing deprivation variables were tested but rejected because they did not offer any significant improvement in the amount of variation in direct measures explained by other variables.

Whether a new variable is relevant can be tested by including it in a new regression against the dependent variable. The new variable improves the statistical model if a significantly higher correlation is established with the dependent variable, i.e. more of the variation in the dependent variable is accounted for by variation in the explanatory variables. However, this is not by any means the sole criterion for selecting indicators. The identification of relevant variables is largely a matter of background research into the type of causal relationships that may be present as well as judgement about what is appropriate.

Although the identification of a client group for a service such as primary education is straightforward, this is not always the case. For example, the client group for under-5 education is the *potential* group of service recipients – the population aged under 5 years in a local authority area. A more complicated example is assessment of the need to spend on personal social services for children, which include residential, foster and nursery care, services for children with disabilities and learning difficulties, and associated social work and administrative costs. For this sub-block, the client group is estimated from selected indicators rather than the number of children receiving personal social services. The latter is not an indicator of need because this will be affected by the level of services actually provided. Instead, the client group is the estimated number of *children at risk*. This is derived using a *children in need index* based on a number of indicators found in regression analysis to be associated statistically with the proportion of children in community, voluntary and foster homes.

The results of regression analysis carried out by the Department of the Environment indicate that the proportion of a local authority's child population that is likely to be at risk and therefore in need of social services care is related to the following variables:

- The proportion of children under 18 living in lone-parent families weighted by ×3 (from the 1991 census).
- The proportion of children under 16 living in rented accommodation (from the 1991 census).
- The proportion of children whose parents claim Income Support (from Department of Social Security returns).
- A measure based on the number of households with children accepted as homeless by the local authority.

The children-in-need index is a composite of these variables which produces for each local authority its estimated number of children at risk (the full details of this calculation are beyond the scope of this book but further details may be found in Department of the Environment, 1994). Note that lone-parent families is given a weight of ×3 relative to the other three variables. This increases the explanatory power of the regression model compared with a lower weighting, reflecting the strong correlation between this variable and children in community, voluntary and foster homes.

Once the number of children at risk has been estimated using the children-in-need index, this can be multiplied by the average annual cost per head of providing social services to children to give the basic spending assessment for these services. However, adjustments are also made to take account of the increased costs associated with particular social conditions, as well as the area cost adjustment. The indicators of social conditions which attract additional funding are the proportion of residents in the local population who share their accommodation and the proportion of non-white children under 18.

In summary, the SSA element for the social services for children sub-block for a particular local authority in 1994/5 is calculated as follows:

1 *Children at risk* multiplied by the result of: £11,227.37 (the basic unit cost allowance); plus the following allowances which reflect additional costs associated with two particular factors: £23,261.82 multiplied by the *shared accommodation factor*; plus £13,986.11 multiplied by the *ethnicity factor*.
2 The result of (1) is multiplied by the *area cost adjustment* for general personal social services.
3 The result of (2) is then multiplied by a *scaling factor* which ensures that the sum of the SSAs for all authorities adds exactly to the government's limit for spending on social services for children (the 'control total').

One block of services known as 'all other services' is an aggregation of all leisure and environmental services, passenger transport, planning and economic development services. There is no obvious client group. An indicator called 'enhanced population' is used to measure the need to spend on this block of services, based on local population adjusted to take into account numbers of commuters and visitors to each local authority area. The formula also includes population density and sparsity factors, a social index and an economic index.

The social index has proved controversial because it fails to reflect the distribution of socio-economic deprivation in the country. The housing indicators in particular produced results which strongly skewed resource allocation towards authorities such as Cambridge, Bristol and Oxford which had a high incidence of students and young people sharing accommodation. Following a major review of SSAs in 1993, the social index was revised to include a measure based on the number of homeless households. More significantly, however, a new economic index was added. This includes variables based on unemployment benefit claims, housing benefit claims, residents in lone-parent families and premature mortality (a measure of ill-health found in statistical analysis to belong to the economic index). The inclusion of an economic index shifted resource allocation for the other services block towards authorities with high rates of poverty. The new economic and social indexes together resulted in metropolitan areas gaining approximately 10 per cent in the 1994/5 RSG settlement compared with simply updating the SSA for the 'other services' block with 1991 census data. London lost just under 5 per cent and the shire areas approximately 2 per cent.

PROBLEMS WITH SSAs

As noted above, SSAs incorporate adjustments which seek to take into account the fact that unit costs vary between local authorities. Examples are adjustments made in the SSA elements for primary and secondary education to reflect the increased costs associated with the additional educational needs of some pupils, the provision of free school meals and the sparsity of the population in a local authority area. The indicators used to make these adjustments are derived from judgement and research about the factors that influence spending, preferably based on evidence of an underlying causal factor such as 'social deprivation'. It must not be possible for indicator variables to be manipulated by local authorities for unfair advantage. Although the indicators and their relative weights are tested using regression analysis, the final formulae used in SSA calculations are ultimately based on political decisions about how resources should be allocated to local authorities.

Most assessments are based on regressions of actual local authority activity levels or actual unit costs. The dependent variable used in the development work on SSAs is often past actual local authority expenditure. As a result, SSAs are significantly influenced by levels of historic spending. It is of course arguable whether this reflects actual needs. The use of 1990/1 expenditure in development work for new SSAs introduced for 1994/5 was strongly criticised by the Association of Metropolitan Authorities for biasing allocations towards smaller, largely Conservative-controlled, districts (*AMA News*, November 1993). This was because spending in metropolitan areas had been suppressed by capping and the introduction of the poll tax (see pp. 84–90). Spending by authorities with budgets less than £15 million was at this time not subject to capping, even though many of these authorities had increased their spending sharply. The assumption that in metropolitan areas the level of spending in 1990/1 is a proxy for the level of need is thus very questionable given that their budgets were subject to limits set by central government. The government response to this criticism was that because of the timing and limited scope of capping in 1990/1, there was only a minor impact on the regressions. Martin (1994), however, shows this not to be the case (see below).

Where they exist, it is clearly preferable to use direct service measures as dependent variables rather than expenditure. For example, in the case of the education block the AMA has suggested that the results of Standard

Table 4.6 'Degree of fit' of original regression models used in SSA calculations

Original regression	*Percentage of variation 'explained'*
Primary/secondary education	47
Other education	63
Children at risk	73
Children's unit cost	41
Elderly residential social services	60
Elderly domiciliary social services	12
Other personal social services	67
Other services (county councils)	60
Other services (district councils)	64

Source: Fairclough, 1992.

Assessment Tasks (SATs) at Key Stage 1 be used as a direct measure of the need to spend on education across local authorities (Association of Metropolitan Authorities, 1993c) However, it is often difficult to identify measures which are not influenced by local policy decisions, political preferences and variations in the efficiency and effectiveness of services.

The explanatory power of a regression model is measured in terms of the amount of variation in the dependent variable which the selected independent variables 'explain'. This figure is calculated by squaring the correlation coefficient derived from the regression equation and is known as R squared. Table 4.6 shows 'adjusted R squares' for the original regression models used in SSA calculations up until 1994/5. Fairclough (1992) uses adjusted R square in this table because R square tends to be overoptimistic about how much variation is explained by the model and therefore an adjustment based on the number of cases and variables tested is made. The explanatory power of the regressions varied from 73 per cent for children at risk to only 12 per cent for domiciliary social services for elderly people. The Department of the Environment's view is that unexplained variations largely represent influences on spending other than need, such as differences in political preferences and levels of efficiency between local authorities.

The elderly domiciliary regression was unusual because data from the General Household Survey was used rather than local authority expenditure. This involved a much larger number of observations (over 3,000 households) than all the other regressions used for SSA purposes. A high amount of unexplained variation is therefore more likely.

Unexplained variation means that the model is not fitting the current pattern of service provision, an issue that is much more serious for authorities which significantly deviate from the established pattern of a relationship between independent and dependent variables calculated from a regression analysis. Since SSAs were introduced, metropolitan districts and inner

Table 4.7 Changes in SSA by type of authority, 1993/4 to 1994/5

Type of authority	*Change in SSA (%)*	*£m*
Shire Districts	+ 5.6	+ 156.2
Shire Counties	+ 2.0	+ 367.8
Metropolitan Districts	+ 1.3	+ 106.9
Metropolitan Police Authorities	+ 4.5	+ 29.7
Metropolitan Fire Authorities	+ 0.6	+ 1.9
Inner London	− 1.2	− 33.0
Outer London	+ 0.7	+ 25.2
Metropolitan Police	+ 3.8	+ 29.6
London Fire and Civil Defence Authority	+ 1.1	+ 2.6
England	+ 1.8	+ 686.9

Source: Department of the Environment (1994a).

London boroughs have tended to spend more than their SSAs, suggesting that their spending needs are underestimated by the formulae used in the SSA calculations.

The Department of the Environment involves local authority associations and other government departments in annual reviews of the methodology, although amendments are kept to a minimum over a number of years to promote stability. A major review undertaken in 1993 was carried out to take into account the replacement of 1981 with 1991 census data and the need to fund councils' new community care responsibilities through the SSA. Table 4.7 shows the changes in SSAs by type of authority which followed this review, with a breakdown to show the effects of (a) different control totals; (b) incorporation of the 1993/4 Standard Transitional Grant for community care into SSAs (see below); (c) the SSA review and incorporation of 1991 census data; and (d) residual data changes, i.e. data items that would have been updated regardless of the 1991 census and review, together with the effect of local authority boundary changes.

Table 4.7 shows how London lost heavily from the SSA review. Further analysis by Martin (1994) indicates that compared with the 1993/4 settlement the effect of the new 1991 census data alone was to increase the total SSAs of the London boroughs by 1.2 per cent and of the metropolitan districts by 0.5 per cent, whilst the shire counties lost 0.7 per cent and the shire districts neither lost nor gained. However, the effect of changing the expenditure base to 1990/1 worked in the opposite direction, decreasing the total SSAs for the London boroughs by 3.4 per cent and the metropolitan districts by 0.5 per cent, but increasing those of the shire counties by 0.7 per cent and the shire districts by 6.1 per cent. The change of the expenditure base from 1987/8 to 1990/1 expenditure had a bigger impact on the settlement distribution than either the new census data or other changes such as in the distribution of spending between services, methodology, or data changes other than the census.

An important outcome of this review was the Department of the Environment's decision to incorporate an *economic index* in the 'other services' block, as noted above. This followed strong lobbying by metropolitan authorities which considered that their levels of unemployment and poverty were not adequately reflected in the assessment for this block. Homelessness and limiting long-term illness (a new census variable in 1991) were also incorporated into the personal social services assessment. However, the proportion of education resources linked to additional educational needs was reduced from 21 to 16 per cent, thus increasing the amount allocated on the basis of actual pupil numbers alone. The Association of County Councils had lobbied for this reduction, arguing that the additional educational needs index allocated to authorities that scored highly on the index more resources than were actually spent on pupils receiving special provision. These authorities could then spend more on other pupils. The Association of Metropolitan Authorities opposed this view, arguing that authorities which scored highly on the index needed additional resources for mainstream provision because of the additional demands caused by the effects of social deprivation and home background on many pupils.

Before these changes, about half of local authorities did not regard SSAs as an acceptable basis for grant distribution (Audit Commission, 1993a). In particular, SSAs were regarded as unfair by urban authorities due to the absence of measures of economic conditions and ill-health which, they argued, forced up spending on services in their areas (Association of Metropolitan Authorities, 1992). The changes have gone some way towards meeting these criticisms. But the small proportion of spending now financed by local taxation, combined with capping, means that local authorities cannot use local taxation to compensate for imperfections in SSAs arising out of their particular local situations.

Allegations of political bias in how SSAs are constructed have been made since their inception. This is understandable when a central government of one political party is allocating money to local councils often controlled by a different political party. Reviews of SSAs involve close scrutiny of which local authorities benefit and which lose out from any changes made. A controversial example was a decision in 1991 to include 'foreign visitor nights' in a measure of enhanced population, thus increasing SSAs in authorities with a high number of such visitors. As a result, in 1992/3 Conservative-controlled Westminster gained £5.3 million and Conservative-controlled Kensington and Chelsea gained £4.5 million in RSG. However, Hackney lost £600,000, Islington lost £700,000 and Newham lost £700,000; all Labour-controlled authorities. One Labour authority, Camden, gained.

SSAs can never be an objective method of distributing resources because there are few generally accepted definitions or measurements of need. Assumptions and judgements have to be made to define the problem and enable any analysis to be undertaken. The statistical methodology used is not an objective analysis but a consequence of the value judgements on which it is based. SSAs are therefore inevitably open to manipulation. Indeed, when SSAs were introduced in England, inner London had a particular problem because spending levels had historically been much in excess of the old grant-related expenditure assessments as a result of income from high non-domestic rates. The power to set non-domestic rates was removed from all councils at the same time that SSAs were introduced, threatening a financial crisis for inner London's services. The problem was addressed by changing the weightings given to social indicators in the SSAs and giving a higher weight to the personal social services and 'other services' blocks. This redistributed grant from other authorities back to inner London, although the changes made were not obviously relevant to the need to provide services (Audit Commission, 1993a).

One solution to the controversy surrounding SSAs that has been suggested is to establish an independent Grants Commission to allocate funding once central government has determined the total resources to be made available. However, in Australia, where this system is in place, it has given rise to complex formulae with large data requirements without eliminating the necessity to take value judgements (Audit Commission, 1993a).

LOCAL TAXATION AND 'CAPPING'

Local taxation to pay for local government services was long based on a property tax called 'the rates'. This dated from when most local government services were services to properties, such as water and sewerage, rather than to people. The higher the value of the property, the more 'rates' the owner paid. This was regarded as progressive (that is, people with more valuable property pay more tax) because there is a broad relationship between the value of people's homes and their incomes.

The ending of rates and the subsequent rise and fall of the 'poll tax' was an extraordinary period of British politics. It brought mass protests and rioting on to the streets, induced Conservative councillors to resign, forced £14 billion of government spending in an effort to save the tax, and was a decisive factor in the downfall of Margaret Thatcher as prime minister.

Public policy must be understood in political terms. Reade (1987) defines the political clarification of public policy to involve the analysis of its distributional consequences – who benefits and who loses. He argues that public policies promote particular interests, even though this may be obscured by not making explicit the distributional effects of decisions. The poll tax is a very good example.

In the three years between a 1983 White Paper and a 1986 Green Paper on local taxation, the government made a dramatic U-turn in reversing the 1983 White Paper's argument that domestic rates promoted local accountability. The catalyst for the rapid reassessment of this view was a crisis of the rating system caused by two factors. First, an impending property revaluation in Scotland which, despite cushioning measures, would shift the rates burden from non-domestic (industrial and commercial) ratepayers to domestic ratepayers. Second, the impact on local rates of a decline during the 1980s in the proportion of local government expenditure funded from central government (rate support grant).

In 1985, the Confederation of Scottish Local Authorities (COSLA) showed that as a result of these two factors domestic rates would have to increase by 21 per cent. One-third of this was due to revaluation and two-thirds to the decline in the proportion of spending met by revenue support grant. The political consequences were massive, as McConnell explains:

> Given that 'desirable' residential areas tended to have the highest revaluations, then the sharpest impact of the shift in the rating burden was felt in areas with Conservative-held seats, particularly Edinburgh, Eastwood, Stirling and Bearsden. Arising out of this, therefore, was a widespread condemnation of the rating system, and demands that *something* be done ... This 'outcry', directed towards the Scottish Conservative Party from its traditional base of support, was translated into demands within the Party that the 'rates must go'.
>
> (McConnell, 1990: 68–9)

The possibility of further electoral losses for the Conservatives in Scotland, where already only 21 of a possible 71 parliamentary seats were held by the Party, would not only be a disaster in Scotland but also threaten the Conservative's claim to be the national party of a unitary state. The prime minister, Margaret Thatcher, had long held a desire to abolish the rates, but no alternative form of local taxation had been regarded as feasible. The Scottish problem, however, made the replacement of rates essential.

Various alternatives were considered during 1985. A poll tax had most support from groups affected by revaluation, but was opposed by organisations ranging from the Labour Party to professional bodies such as the Chartered Institute of Public Finance and Accountancy (CIPFA). Opposition centred on a poll tax being regressive (it would hit low income groups hardest), impractical and a further undermining of the power of local authorities to spend according to local needs rather than to central government prescription. In the end, the government came down in favour of the poll tax because it meant that each voter would bear a fixed share of any increased tax costs arising from their council's spending. Critics of the tax were labelled as supporters of local government 'overspending' who would deny 'fairness' to ratepayers. The new tax was introduced into Scotland in April 1989 and England and Wales in April 1990. It was not introduced in Northern Ireland. Its official title was the 'community charge', a term that was rarely used.

The principle that every voter should pay the poll tax was extended to people on social security benefits and low wages who, under the old rates system, could receive rebates of up to 100 per cent. Benefit claimants had to pay 20 per cent of their poll tax bill. This proved very costly to collect and many benefit claimants could not or would not pay their 20 per cent contribution. In inner city areas, up to 60 per cent of the entries on a poll tax register were likely to change in the course of one year. The Audit Commission estimated that collecting the 20 per cent contributions cost £15 in administration for every £6 of net revenue it raised.

The poll tax reduced the local tax burden on people living in the south-east of England in high-value and smaller households. It increased the burden on lower value homes and households with three or more adults in the northern regions of the country. At district council level, affluent areas had more people gain than lose from the introduction of the poll tax. Conversely, in poorer areas more people lost than gained from the new tax (Newcastle City Council, 1986).

The political impact of the tax for the Conservatives was in fact disastrous. This is because many voters in key *marginal* constituencies saw their local tax burden rise.

The poll tax was extremely unpopular and many people did not pay it. Many people sought to evade the tax by not registering to vote. The consequence of non-payment was that 1992/3 bills included a 'non-collection' surcharge of on average £20. By the end of 1991/2, local councils in England had issued 11 million summonses for non-payment and had at least £1.5 billion outstanding (*Hansard*, 1992e). It is estimated that chasing and collecting outstanding poll tax during 1993/4 cost councils some £250 million, the first year of the new council tax, and some authorities may still be collecting arrears beyond the year 2000. In Scotland, local councils face arrears of more than £620 million.

In order to reduce the impact of the poll tax, the government reduced its level by increasing value added tax from 15 per cent to 17.5 per cent from 1991/2, raising £4.25 billion. This substituted national taxation for local taxation, reducing the contribution of the poll tax from 34 per cent of total spending in 1990/1 to 22 per cent in 1991/2. As a result, the poll tax in England was reduced from an average of £392 to £252.

Protests against the unfairness of the poll tax continued, however, and the government was eventually forced to replace it. Five principles were established to justify the introduction of its successor, the *council tax* (*Hansard*, 1991, col. 404). These were:

1 People should be able to see a link between what they are paying in local taxation and what the council is spending – that is, accountability to taxpayers.

2 To be acceptable, taxes have to be perceived to be fair.
3 The tax should not be too difficult to collect.
4 Most people should make some contribution. Specifically, any tax should take some account of the number of adults in each household.
5 No one should have to pay 'penal' rates of tax. One of the Conservative's major criticisms of the old rates system is that a minority of electors paid very high bills.

The council tax replaced the poll tax from 1 April 1993. The uniform business rate remained unchanged. The council tax is partly a property tax and partly a personal tax, with one bill per household. The tax is levied on houses and flats according to their market value as assessed for 1 April 1991. Properties are placed in one of eight capital value bands. In England, these range from band A (value up to £40,000) to band H (£320,000 and above). There are different values for these bands in Wales and Scotland.

A system of discounts and benefits operates. Single person households pay only 75 per cent of the tax for their property band. People liable for council tax and receiving income support have all their council tax met by council tax benefit, in contrast to the poll tax which required everyone to pay at least 20 per cent regardless of their income. People on low incomes pay a proportion of the tax according to a taper. People with disabilities may qualify for Disabled Relief which reduces their tax liability to the band immediately below the one in which their property is placed (no relief is given for a band A property). Students are exempt. Thus, the tax has features of both the old domestic rates and the poll tax because it has both a property and a 'head count' element, but seeks to avoid the worst characteristics of both.

Local councils officially set their council tax for the fourth band, band D. Those in the other seven bands pay a proportion of the tax for band D properties. This proportion is set by central government. Presently, those in the highest band pay twice the central band D amount, while those in the lowest band pay two-thirds.

Table 4.8 shows for illustrative purposes how the band D council tax was calculated in the case of Newcastle City Council for 1993/4. The council's SSA was £204.8 million. If all of this was to be raised by the council tax, a band D property would have had a tax of £2,654. However, the council received £169.8 million of external support from government grants and business rates, representing 83 per cent of its SSA. This means that if the council was to have spent at the level of its SSA, the band D council tax would have been £453.41. The council decided to spend above its SSA, although only so much as to keep it within the government's capping limit. The effect of this was to add £230 to the band D council tax, making a total tax of £684. On to this had to be added a deficit on the old poll tax Collection Fund, adding a further £49 to the tax bill. Two 'precepting' bodies raised the band D council tax to £791.81. Precepting bodies include county councils, which

Table 4.8 Calculation of band D council tax: Newcastle City Council, 1993/4

	1993/4 (£)	*Band D (£)*
City Council		
Expenditure at SSA	204,808,000	2,653.98
Less: external support	(169,818,000)	(2,200.57)
expenditure above SSA	17,762,000	230.17
	52,752,000	683.58
Plus estimated deficit on collection fund 31.3.93	3,800,000	49.24
	56,552,000	732.82
Northumbria Police	3,226,000	41.81
Tyne and Wear Fire and Civil Defence	1,326,000	17.18
	61,104,000	791.81

precept district councils in areas with two-tier local government, as well as police and fire authorities, their requirements for funds from council taxpayers being collected on their behalf by the billing authority. Parish councils are also precepting authorities. Many individuals do not pay the full council tax because of entitlement to single-person discounts, council tax transitional relief, and rebates.

Local authorities which have most of their properties in the higher council tax bands now receive smaller revenue support grants than under the poll tax. Thus, the tax has partly reversed the shift of resources which took place in 1990 when the poll tax replaced domestic rates. However, the council tax is not as progressive as the old rates system because instead of the tax liability simply rising with an increase in the property value, properties in the top and bottom valuation bands receive flat-rate tax bills:

> Band A and Band H are catch-all valuations. Band A in England will apply to all properties worth £40,000 or less (£30,000 in Wales, £27,000 in Scotland), while Band H will cover all those worth over £320,000 (£240,000 in Wales, £212,000 in Scotland). However small or appalling a property, it cannot have a valuation lower than Band A. Equally, Band H will contain everything from substantial private homes to castles.
>
> (Travers, 1992: 29)

For 1993/4, the average council tax bill was £444, 11 per cent lower than 1992/3's average poll tax bill for a two-person household. However, there were wide variations across the country, including large increases over the poll tax in Conservative-voting parts of the south of England. In the first year, 1993/4, the increases were cushioned by the council tax transitional reduction scheme if they exceeded certain thresholds. The amount of this relief was reduced for 1994/5, and at the time of writing no decision has been made about whether relief will be continued in future years.

The most damaging thing about the poll tax for the Conservatives was its impact on marginal voters. The council tax has lifted some of the burden off this group, to the cost of people in higher-value properties. It is tempting to surmise that the government believed that this group will remain faithful to the Conservative Party in their voting behaviour.

Some commentators have predicted that the council tax is unlikely to be a long-term and durable replacement for the poll tax (McLeod, 1992). One of the main problems is that central government grant to local authorities has to be kept at a high proportion of total local government spending in order to keep the 'headline' rate of council tax down. This intensifies the gearing effect whereby local taxes rise faster than local spending. The solution seems to be either to find a more acceptable tax base for local government or for local government to become a local administrative arm of central government.

Alternatives to the council tax include a local income tax collected by the Inland Revenue, which is favoured by the Liberal Democrats. The Conservative Party opposes this because it does not wish to give large numbers of Labour councils the power to set levels of income tax, and is particularly concerned that a minority of high earners would be taxed heavily, yet be unable to control the outcome of local elections. A local income tax would also infringe the principle of a specific local tax for local services.

The Labour Party's policy is 'fair rates'. These would be a property tax based on a combination of market values, rebuilding costs (excluding land values), and maintenance and repair costs. Rebates for low-income households would be up to 100 per cent and there would be generous allowances for people living alone. The Party has no plans to increase the proportion of council spending raised by local domestic taxation. Labour would replace the uniform business rate with locally set business rates, but with increases having to be linked to increases in fair rates. A Labour government would not cap the level of local rates; instead, annual elections would be expected to maintain accountability to local taxpayers and guidelines would be issued on the amount of council spending which will be supported through central government grants.

Under present Conservative legislation, local authorities have the freedom to set the council tax, but this is severely limited by 'capping' how much a council is allowed to spend. Capping has two main purposes. It is used to control total local authority spending and thus the share of national income taken by this spending; and it is used to 'protect' local taxpayers from bills which central government regards as excessive.

The capping of local taxation was first introduced in Scotland in 1981. This ended the principle that local councils should determine their own levels of

Table 4.9 Criteria for limiting council tax in 1994/5

*Increase/decrease in budget compared with 1993/4**	*Council tax capped if results in:*
+1.75% or more	Budget above SSA for 1994/5
+1.25% or more	Budget over 5% above SSA
+0.75% or more	Budget over 10% above SSA
More than 0	Budget over 12.5% above SSA
0	Budget over 30% above SSA
–5% or more	Budget over 60% above SSA
–10% or more	Not capped

Note: *Includes adjustments for community care special grant and any population loss grant. Any costs arising from local government reorganisation are ignored.

expenditure as long as this was financed from their own local taxes and subject to local electoral mandates. Instead, central government could intervene to fix the rates of councils whose budget expenditure was deemed to be 'excessive and unreasonable'. The legislation was extended to England and Wales by the 1984 Rates Act, despite widespread opposition (Duncan and Goodwin, 1988).

In 1993/4, provisional capping *criteria* were issued for the first time by the Secretary of State for the Environment to enable local authorities to avoid setting budgets which would result in capping. The capping criteria are a combination of the extent to which an authority plans to spend above or below its SSA *and* the extent to which the authority intends to increase or decrease its budget compared with the previous year (see Table 4.9). Thus, capping takes account of both 'excessive budget requirements' (spending above SSA) and 'excessive increases in budget requirements' (increases compared with the previous year). The government considers that there should be smaller increases in an authority's budget requirement for those authorities budgeting higher relative to their SSA.

In 1993/4, for urban authorities the resulting maximum budgets were very tough. Forty-six of the 69 metropolitan districts and London boroughs were permitted a year-on-year increase in spending of 1 per cent or less. Budgets were also kept very close to SSAs. Thirty-three out of the 69 districts and boroughs could not spend more than 2 per cent above their SSA without exceeding the provisional capping criteria.

COMMUNITY CARE

A new policy of community care came into effect from 1 April 1993 and is a major area of responsibility for local government. Urban areas have a wide range of community care needs. Inner cities have large populations of older people who are the most important user group for community care services.

Frailty, disability and mental illness/dementia increase with age and demand responses ranging from home helps or adaptations to houses and flats, to full-time care in a residential or nursing home.

The aim of community care is to help people with social care needs to lead their lives as independently as possible, preferably in their own home. But it is widely claimed that the resources available are insufficient. The Association of Metropolitan Authorities estimated that nationally funding for community care in 1993/4 was £54 million less than needed to maintain services at their 1992/3 level.

The main financial change from April 1993 affects new admissions of people needing publicly funded residential or nursing home care provided by the private or voluntary sector. This used to be provided by the Department of Social Security (DSS) paying a higher level of Income Support to the person who needs this level of social or nursing care. From April 1993, responsibility for paying for new admissions passed to local authority social services departments. This financial help depends on whether the person is assessed as in need of residential or nursing home care. The assessment may indicate that the person will be able to live in their own home with a suitable package of domiciliary services and respite care.

Income Support payments before the April 1993 changes had fuelled a very large growth of private sector residential and nursing homes during the previous decade. Over the period from 1987/8 to 1990/1 there was an average annual growth in each financial year of between 20,000 to 25,000 occupied beds nationally in both private and voluntary homes. Local authorities are expected to continue using this 'independent' sector to provide care.

With the transfer of responsibility for paying for beds from the DSS to local authorities, the Department of Health has to estimate how much money to transfer to local authorities in future years. This is a complicated task which has to take into account such factors as faster throughput in acute hospital beds and demographic pressures. In contrast to the old system of DSS funding, local authority budgets for care are *cash-limited.*

In the first four years of the new community care policy, 1993/4 to 1996/7, special government grant for community care services is 'ring fenced' to ensure that local authorities spend the money specifically on community care. This is known as 'special transitional grant' (STG). It is equivalent to an estimate of what would have been spent by the Department of Social Security in Income Support payments to people placed in residential and nursing homes, although the new policy is encouraging a diversion of people with care needs from residential homes to day and domiciliary care. Additional amounts were included in 1993/4: £26 million transferred from the Independent Living Fund (ILF) which the government decided to close, and £140 million to finance the development of contracting and purchasing functions, and the introduction of new information technology and financial systems. STG totalled £565 million in 1993/4. Local authorities are required to spend

85 per cent of their allocations on purchasing care from the private and voluntary sectors.

In 1993/4, STG was not allocated entirely on the basis of SSAs. Instead, the basis for allocation was that half of the money was distributed according to the geographical pattern of Income Support expenditure on care in private and voluntary sector residential homes and half through the SSA. However, urban authorities with high social needs have fewer private sector homes. They also tend to be net exporters of people into care in other local authority areas. Thus, the effect of basing half of the allocation on private sector provision was to penalise authorities with greater needs. Nevertheless, to transfer the DSS funding entirely on the basis of the SSA would have resulted in a marked shift of money away from the present pattern of independent sector provision, causing major disruption in the sector. The government therefore decided to give the private sector time to adjust to the new system.

The part-Income Support, part-SSA distribution was only used for the first year of the STG. In 1994/5, the STG for 1993/4 was transferred into the SSA, excluding the ILF element. Thus, except for the ILF element, this first STG remains in the system but for 1994/5 is distributed entirely by SSAs as revenue support grant. A second STG, totalling £736 million, also had to be distributed to meet the cost of the estimated additional needs for residential and domiciliary care arising in 1994/5 that would have previously been met by Income Support. This total also included £64 million to meet the cumulative resource consequences of closing the old Independent Living Fund and £20 million to encourage the development of new home care and respite care services.

Thus, for 1994/5 and subsequent years the STG will be distributed entirely on the basis of the SSA, although it will remain ring-fenced. The effect of this is to re-distribute spending on community care previously met by Income Support from the shire districts to the metropolitan districts and London.

FUNDING SCHOOLS

Since 1990, state schools have had budgets gradually delegated to them from their local authorities under the Local Management of Schools (LMS) initiative. LMS has given school governing bodies and head teachers considerable discretion over how the school's money is spent.

The potential schools budget (PSB) comprises a local authority's total spending on its schools minus the cost of capital expenditure and interest payments, home-to-school transport, school meals, the council contribution to specific government grants, and certain 'exceptions'. Councils must delegate at least 85 per cent of the PSB to individual schools, a proportion that has been gradually increased and is likely to be further increased in the future. This is called the Aggregated Schools Budget (ASB). It is distributed to individual schools using a formula largely based on the number and age of pupils. Local authorities have some discretion to adjust it, such as taking into account deprivation by weighting the formula using the number of pupils receiving free school meals. However, the details of the formula have to be approved by central government.

Schools can choose to opt out of the control of their local authority and become grant-maintained (GM), funded directly by central government through the Funding Agency for Schools. A majority of parents must vote in a ballot to support such a decision. The government deducts the grant it pays to a GM school from the Revenue Support Grant it pays to the council. The GM grant comprises the budget that the school would have had under the local authority's LMS formula plus an amount called the 'standard add on' that covers the costs that were previously met centrally by the council from the retained proportion (up to 15 per cent) of the general schools budget. If more than 15 per cent of the primary or secondary schools in a local authority area become grant-maintained, the 'standard add on' can be replaced at the request of the school or the local authority by an amount equivalent to the actual expenditure the authority would have incurred centrally for the schools. The GM grant also includes a subsidy to pay for pupils claiming free school meals plus any extra subsidy that the council provides towards the cost of meals for other pupils.

In 1994/5, the government introduced a Common Funding Formula (CFF) in five local education authorities, these councils' own LMS formulas having until then formed the basis for funding GM schools in their areas (Department for Education, 1993a). The CFF is based on four core components: age-weighted pupil numbers (which distributed over 80 per cent of funding), a block allocation for schools' fixed costs, special educational needs, and pupils receiving free school meals. The amount of money to be distributed via the CFF was based on the education SSA for each authority. To derive a figure for secondary schools alone, each local authority's latest financial information was used to deduct from its total education SSA the proportion spent on continuing local education authority responsibilities, such as services to pupils with special needs, and the proportion spent on primary schools. The remainder was then allocated by the formula across the authority's secondary schools. For LEA schools, this was a notional budget, but for GM schools it was an actual budget allocation. Allocations were adjusted to ensure stability in each school's funding. In addition, if the authority spent above SSA on its secondary schools, this was also applied to GM schools to ensure that they were not disadvantaged. Similarly, the CFF means that in underspending local authorities spending at the secondary SSA for GM schools will be phased in. In 1995/6, the government plans to extend the CFF to local education authorities where more than 30 per cent of secondary pupils are in GM schools, which would presently apply to a further sixteen areas (Department for Education, 1994). The CFF means that central government more directly controls spending on

GM schools, rather than this depending on council decisions about how much to allocate to education services from their overall budget (Wills, 1994).

FUNDING SOCIAL HOUSING

Social housing for rent is provided by housing associations and local councils. Housing associations are now the main providers. In 1992, housing associations completed just under 26,000 homes in the UK, compared to 5,400 local authority completions. This very low output of new social housing has resulted in an estimated shortage of about 100,000 homes each year over and above those built by private builders for sale (*AMA News*, April 1994). Local councils remain by far the largest social landlord, with some 87 per cent of the existing social housing stock as a result of their major provider role in the past. There are still four million council tenants compared with one million housing association tenants.

From April 1996, the management of council housing will be subject to compulsory competitive tendering. Local councils' own housing management organisations will have to compete with other organisations such as housing associations and private companies for contracts to manage the stock. Some councils have started letting management contracts in advance of CCT.

Housing associations are Conservative governments' preferred method of providing social housing for rent. In recent years, about 30 per cent of their funding has come from the private sector, with about 70 per cent from the government. However, government support is being reduced and will decline to 55 per cent by 1996, with housing associations expected to raise private finance to meet the difference. The result will be higher rents. The National Federation of Housing Associations has estimated that average rents for new tenants of housing associations who are in work will rise from £48 a week to over £84 by 1996 (*The Guardian*, 23 April 1993). This will threaten the affordability of housing association properties for those low-income households who are not eligible for full housing benefit to meet their rent payments. About half of housing association tenants receive housing benefit either because they are unwaged or are on low-wages. This proportion is expected to rise from 47 per cent in 1992 to 66 per cent in 1996 as a result of rent increases.

The increasing dependence of tenants in social rented housing on housing benefit continues the trend begun in 1980 of reducing government support to housing investment and redirecting subsidies to individuals receiving housing benefit. A particular problem with this strategy is the poverty trap, especially when high rents result in tenants who are in work having to claim this benefit. Every additional pound earned may be largely lost as a result of subsequent reductions in entitlement to housing benefit and other means-tested benefits such as family credit. This can make it largely pointless to work or to earn more money; taking into account transport and child-care costs, it may be more sensible for a low-paid tenant to give up employment.

The housing revenue accounts (HRAs) of local councils are 'ring fenced', keeping housing finances separate from other council finances. The rents paid by tenants meet the *revenue costs* of managing and maintaining the stock and paying debt charges on money borrowed for building and refurbishment. The low incomes of tenants of both local authorities and housing associations mean that most of this rental income is in fact from housing benefit. Whilst central government subsidy of revenue costs is now very small, with many authorities receiving no subsidy or making surpluses on their HRAs, the government meets a high proportion (over 90 per cent) of the cost of paying housing benefit.

Local authority and housing association *investment* is co-ordinated at local level within local authority's annual housing strategy statements. Thus, all public sector resources are applied to addressing the needs of a local authority area in a co-ordinated way.

Housing strategy statements were first introduced in 1993. They describe the needs and resources for housing in the local authority's area, and the plans and programmes which the authority has for addressing them. The statements are required to identify clear targets and to monitor performance against these targets each year. They are developed in consultation with the Housing Corporation, housing associations, the private housing sector (organisations in both the rental and owner-occupied sectors), house builders, the voluntary sector, tenants' groups and other interested parties. Central government's housing policies should be reflected in the statements and their emphasis should be corporate, encompassing all relevant functions and resources. This includes the role and contribution of the private sector and how this will be enabled; the involvement of housebuilders in projects funded by sources such as City Grant, for example, has become increasingly common, developing refurbished or new housing for both sale and rent. Another aspect of the corporate approach expected in housing strategy statements is liaison with social services regarding care in the community and children in need.

Housing strategy statements accompany the annual local authority housing investment programmes (HIPs), although they are separate documents. HIPs are bids for borrowing approvals from the Department of the Environment based on the need for investment identified in the strategy statement.

At the national level, central government decides the total of capital investment which there will be in social housing. In 1994/5, HIP allocations to local authorities totalled £1.5 billion and the Housing Corporation's approved development programme for housing associations amounted to over £1.5 billion (Hansard, 1993). Within this, the broad priorities to be given to different types of expenditure are also decided. Local authorities bid for allocations each autumn through the annual HIP process, receiving their allocations as borrowing approvals for the forthcoming year.

Since the mid-1980s, government policy has been that investment in council housing should be directed at modernising and refurbishing the existing stock rather than building new houses. Local authorities also provide means-tested renovation grants to tackle problems of disrepair and unfitness in the private sector and many support housing associations by providing discounted land or grants. Spending, however, is well below the level needed. There is currently a backlog of £10 billion of work necessary to tackle disrepair in the council housing stock of four million properties, and a similar amount is necessary to improve over 1.6 million sub-standard dwellings in the private sector (*AMA News*, April 1994).

If present policies continue, by the mid-1990s almost no new council housing will be under construction as a result of government cuts in borrowing allocations. The sale of the better quality housing stock has often left behind unpopular or defective housing, compounded by social problems in such areas of unemployment and crime. Council rents rose in real terms by 135 per cent between 1979 and 1994, and further rises above the rate of inflation are in prospect. These factors mean that people in employment have largely left council housing. Those who remain and live in the most rundown parts of the stock have experienced decentralised intensive housing management strategies which began to be introduced in the 1980s (see Chapter 6). These have been promoted through Estate Action, a national scheme which involves councils bidding to the Department of the Environment for government borrowing approvals (Estate Action HIPs) to target the refurbishment of rundown estates. Bids have to demonstrate a commitment to involve tenants, decentralise management and partner with the private sector or housing associations. Councils are encouraged to dispose of improved stock to these new landlords or into owner occupation. No new government money is involved; Estate Action HIP approvals are approvals for the council to borrow from private sources in the normal way. They have become increasingly important and a number of councils receive over half of their housing investment allocation from Estate Action HIPs.

It was noted above that the capital spending power of a local authority is controlled by central government through Basic and Supplementary Credit Approvals. The HIP process is used by the DoE for making allocations for housing and the housing Annual Capital Guidelines are a substantial component in the calculation of Basic Credit Approvals. In assessing a local authority's need to borrow money for all capital programmes, the government takes into account the proportion of receipts from the sale of council housing and other assets that it can spend on capital projects. This is limited to 25 per cent of receipts from the sale of council housing. Local authorities have flexibility in how they spend the usable proportion of capital receipts;

thus not all the usable receipts from the sale of council housing have to be allocated to the housing capital programme. However, since the 1980 Housing Act, which introduced the 'right to buy' for council tenants, capital receipts from these sales have increasingly become the normal method for financing councils' housing capital programmes.

In recent years the government has used its control over capital spending to cut it back sharply. Local authority spending on all capital projects fell by 22 per cent in real terms between 1984 and 1994. Housing has been the biggest loser with spending falling by 81 per cent in real terms (*Municipal Journal*, 7–13 January 1994). Local councils now face difficult choices about how to spend capital. For example, should spending be used to bring empty council housing back into use to remove homeless people from bed and breakfast accommodation? Or is a higher priority providing mandatory grants to private landlords to minimise the fire risks of houses in multiple occupation which are frequently occupied by poor or vulnerable tenants?

The relative need for housing investment in an area is assessed by the Department of the Environment using a formula called the generalised needs index (GNI). This is used to assist with distributing HIP borrowing allocations to local councils. These allocations determine the size of borrowing a local council is permitted to undertake to fund refurbishment of the housing stock and new building. A separate index is used in a similar fashion for guideline allocations to local authorities of central government grant for private sector housing renewal.

The GNI is a weighted index of eleven measures of housing deprivation. The indicators are for three groups of need: new housing provision; renovation of council housing; and renovation of private housing. Each indicator is given a weight which is its share in determining the overall GNI and therefore a measure of its relative importance. The indicators are derived from different sources of data, including local authorities' HIP statements, the Labour Force Survey, the Family Expenditure Survey, the DoE House Condition Surveys and the Census.

The GNI is used to make resource allocations to regions, adjusted to reflect differences between the regions in construction costs. It is reviewed and updated on an annual basis in consultation with local authority associations. These reviews have resulted in changes in indicators and their shares. In 1992/3, changes in the GNI resulted in resources being switched away from metropolitan authorities outside London towards London and the southern district councils. The main change was to give greater weighting to homelessness (which rose from a 10 per cent to a 25 per cent share of GNI). The worst affected region was the Northern Region which saw its 'regional' HIP allocation reduced by 11 per cent.

The influence of formula funding has waned in recent years as DoE regional controllers have been steadily given more discretionary influence over the allocations. Although allocations from the DoE to regional controllers are determined by formula, by 1992/3 only 40 per cent of individual local authorities' mainstream allocations from the regions were distributed on the basis of the GNI. Sixty per cent was distributed on a discretionary basis, taking into account existing commitments, the 'quality' of the council's housing strategy and its conformity with government policy, including the extent to which the authority is likely to use its allocation in co-operation with housing associations and the private sector. The DoE publishes regional 'league tables' of the best-performing local authorities, assessed in terms of both the management of their stock and their 'enabling' role.

In May 1992, the DoE issued a consultation paper on the future of the GNI and the HIP process. This proposed abolishing the GNI. Instead, DoE regional controllers' judgements about the relative performance of local councils and the quality of their proposals would have prime importance. This approach is one which Conservative governments clearly favour above allocations based on needs alone. However, the proposal met with strong objections from the local authority associations and was not implemented.

Allocations to housing associations are based on a similar principle using the housing needs index (HNI). Housing associations receive their capital finance from three government-funded agencies: the Housing Corporation (in England), Scottish Homes, and Housing for Wales. Allocations are made as grants which reduce the amount of borrowing which has to be repaid from rental income. A significant proportion of finance is now required to be raised from the private sector. In addition, housing associations have increasingly been used by central government to develop housing for low-cost home ownership rather than for rent, reducing pressure on the limited stock of social rented housing by encouraging those who can afford it to become owner-occupiers. Eighteen per cent of Housing Corporation development funding was for owner-occupation schemes in 1993/4, and this is planned to rise to 28 per cent by 1995/6.

In 1985, the Housing Corporation and the National Federation of Housing Associations jointly developed the HNI as a method for allocating loan finance and subsidy to housing associations. It was reviewed in 1989 and a revised system phased in. By 1992/3, regional allocations were fully determined by the revised system of formula funding. Eighty per cent of housing association grant to local authority areas is on the basis of an HNI score calculated for the area.

The HNI is a combination of indicators derived from different variables. Each indicator is given a weighting because some indicators are judged to be more important than others. The government can modify how the formula distributes money by adjusting the weightings of the various elements of the indicators. Table 4.10 shows the HNI indicators used in 1993/4 and the weights or indicator shares attached to them. Private sector condition, for example, is almost twice as important as needs of the elderly or homelessness.

Table 4.10 HNI indicators 1993/4

Indicator	*Share (%)*
Full household dwelling balance	8
Concealed and sharing households	8
Overcrowding	17
Homelessness	14
Access to owner occupation	4
Needs of the elderly	14
Needs of disabled people	4
Private sector stock condition	27
Numbers in temporary accommodation	4

Source: Lautman and Stearn, 1992.

The HNI determines 80 per cent of the allocations; the remainder is distributed at the discretion of the Housing Corporation's regional directors. This is based mainly on the performance of local councils in giving land or grant to housing associations and the value for money of projects submitted by housing associations. Local authorities with high levels of deprivation have their scores increased to reflect greater priority in these districts.

The total amount of Housing Corporation allocations for distribution to the regions is reduced to pay for special projects such as City Challenge ('top-slicing').

CONCLUSION

Public sector spending has been tightly controlled under Conservative governments. Local authority spending has been particularly constrained, despite requirements to implement major new legislative changes such as the Education Reform Act 1988, the Children Act 1989, the Police and Criminal Evidence Act 1984 and the community care reforms. Important influences on spending, such as teachers' pay awards, are outside the control of local authorities because teachers' pay is now centralised.

There are great pressures on local authorities for increased spending. Growing numbers of very old people are increasing the need for social care. Environmental concerns need to be addressed. Infrastructure and public buildings such as housing and schools provided to poor standards in the 1960s need to be restored. Public expectations of local authority services are higher.

Because the council tax still accounts for only a small proportion of local government spending (some 15 to 20 per cent), the problem of 'gearing' remains, so that big council tax increases become necessary to pay for relatively small increases in spending. The major constraint, however, is central government capping of local authority budgets, which is forcing many local authorities to reduce services and employees to stay within capping limits.

Areas of expenditure which are discretionary – that is, for which there is no statutory duty for the local authority to provide the service – are particularly vulnerable to cuts. Nursery education is a prime example, and has seen cut-backs in recent years despite clear evidence of its importance in providing all children with a good foundation for learning (National Commission on Education, 1993). However, local authorities still act to protect their communities in areas outside their statutory functions. Kent County Council, for example, set aside £2 million to protect local residents from the environmental impact of the proposed high-speed rail link to the Channel Tunnel when this was first mooted, together with measures to secure economic benefits for the county (Brooke, 1992).

As well as this tight central control of local spending, the 1990s have seen central government require local councils to compete for a growing proportion of their resources rather than receive allocations based on assessments of local need. The main considerations which the Department of the Environment appears to use in these competitions are whether local authorities will spend the money effectively and in accordance with central government policy priorities (see Chapter 2).

Two crucial questions emerge from this chapter's discussion of how local public services are funded:

1. Where should the balance lie between central government and local government regarding service provision, financial and policy discretion? The principle of subsidiarity implies increasing local government's responsibilities and freedom to levy taxes to fund local decisions on services. But central government will be concerned that such discretion reduces its ability to control public spending.
2. How far should resources be distributed by formula as opposed to political discretion? Statistical formulae are imprecise and can be insensitive to local circumstances. But giving ministers discretion to allocate resources to local authorities risks claims of partiality.

Part II

APPROACHES TO URBAN POLICY

5

MANAGEMENT OF QUALITY

This chapter is a survey of the range of initiatives that now comprise the management of quality in public services. This subject is often neglected in policy studies, which is surprising given the 'people-centredness' of public policy. The quality of services has to be managed, and providing for the welfare of urban residents has to take into account how services are delivered as well as their levels of funding and policy objectives.

The chapter considers decentralisation and public participation, quality assurance and customer care, citizens' charters and contracting, and policy performance. The chapter ends with an examination of business planning in the public sector and the implications of a quality approach for organisational structures.

THE RISE OF CONSUMERISM IN PUBLIC SERVICES

Public services can vary in quality as well as availability. The quality of services is influenced by policy, spending levels and staff training. Although managers are concerned with all the stages in the production of a service, evaluation of its quality appears to be a question of results for the user at the point of delivery. These results can often be quantified. For example, they could be examination results following schooling, reductions in reported crime as a result of policing, or satisfaction with a housing repair.

The definition of users, however, is not always straightforward. Some users of public services are compelled to receive the service, such as juvenile offenders. People who are not users of a service may none the less have an interest in it and benefit from it. Examples include effective child protection services, which may reduce the risk of future crime and delinquency with its associated social and financial costs, and education, which provides benefits to industry, government and society as a whole, as well as to students. For this reason, 'quality' in a public service is best interpreted in terms of achievement against objectives and standards for the service. These objectives and standards will often have to reflect the needs of consumers at several levels, but achievement against them should be measurable and this should be as important in monitoring services as cost and efficiency.

During the 1980s, public sector organisations were pressed to reduce their costs and improve efficiency. Questions of outcome for the recipients of services received relatively little attention in government policy. This tended to give rise to a preoccupation among managers with financial and volume indicators. Gaster identifies the problems with this narrow focus in local government:

> The result is that, while senior officers, elected Members and district auditors may feel satisfied that targets are being achieved, and while these targets – fewer empty properties, higher proportion of local taxes collected, cheaper meals on wheels – may be objectively desirable in terms of economy and efficiency, they reflect neither the quality of the service on offer, nor the need for that service (compared with other possible services) in the eyes of the users and front-line staff.
>
> (Gaster, 1992: 56)

However, by the end of the 1980s 'quality' – defined in terms of consumer satisfaction – was becoming a major consideration for public service managers who had previously placed greater emphasis on professional judgement. Councillor David Sullivan, when chair of the London Borough of Lewisham's Environmental Services Committee, introduced his department's 'customer contracts' in terms that sum up this new emphasis:

> People in Lewisham deserve high quality services at a cost that they can afford. They must be services that people need, want and value. The services must be provided in a way which suits the customers rather than in a way which suits the organisation.
>
> (Lewisham Environmental Services, 1991: 1)

A number of factors have been behind this adoption of 'customer care' in the public sector. From the middle of the 1950s to the middle of the 1970s, the public sector was based on a bureaucratic model. It enjoyed expanding and stable revenues. Housing, leisure, social care and other key sectors were municipalised on a large scale, and the number of public sector employees grew substantially. The dominant organisational model was large hierarchical departments which mass-produced public services.

During the 1960s and 1970s there was growing public dissatisfaction with this way of managing and delivering services. As Hambleton explains:

> This discontent rolled together concern about the remoteness of centralised decision making, irritation with the insensitivity and lack of accountability of at least some officers, and frustration with the blinkered approach often associated with highly departmentalised organisations.
>
> (Hambleton, 1992: 5)

Consumerism was spreading from the private sector to public services. Paternalistic attitudes to the customer had been common in the private sector until recession in the early 1970s intensified competition and saw 'customer care' develop as a way of increasing market shares and profits (Kelly, 1992). By the beginning of the 1980s, Peters and Waterman (1982) identified 'closeness to the customer' as the hallmark of excellence and innovation in private companies.

In the public sector, there was opposition from individual consumers and community groups to people having their needs categorised by professional experts applying crude aggregate notions of 'problem groups' such as 'the elderly' or 'slum dwellers' (see Chapter 1). By the late 1970s this greater consumer awareness began to influence local authorities, some of which started to adopt strategies of getting closer to their 'customers', principally by decentralising staff from the town hall to neighbourhood offices. But it was the attacks on public sector bureaucracy made by a new Conservative government in the early 1980s that were the major stimulus to replace old bureaucratic ways with a greater responsiveness to service users. Many local authorities responded to government policies of privatisation and competition by highlighting their services as both responsive to needs and democratically accountable.

Several local authorities led the way in applying customer-care ideas to public services in the late 1970s and early 1980s. Walsall Council pioneered bringing services closer to residents by decentralising some services to neighbourhood offices, and by 1990 over fifty local councils had decentralisation schemes (Kelly, 1992). One Stop Shops were also introduced to prevent people having to call in at different offices, giving a single access point to council services. Westminster City Council and Norwich City Council were early examples. In central government departments and executive agencies the 'Next Steps' initiative emphasised enhancing the quality of service delivery and the creation of a 'service culture' (see Chapter 3).

These changes were not wholesale imports from the private sector. A public service orientation marked the various initiatives, especially in local government. The best examples sought to involve users as active participants in setting standards and assessing value, rather than as the passive recipients of services. Decentralisation of some services to neighbourhood offices has often had the aim of building up more positive relationships between service providers and users. Many local authorities have regarded decentralisation as a form of 'empowerment', not only relocating some services out of the town hall but, more fundamentally, establishing local committees and forums as a channel for local views about services. Decentralisation has also included devolving management decisions and budgets to local level, giving greater flexibility to respond to local needs and views.

FROM DECENTRALISATION TO CUSTOMER MANAGEMENT

Decentralisation in the public sector is most in evidence in local government, such as decentralised housing offices, social services offices and planning offices. It was probably the key trend in local government management during the 1980s (Hambleton, 1992).

Decentralisation can take several forms. The London Borough of Tower Hamlets had one of the most extreme models when it was under Liberal Democrat control, with seven neighbourhood committees which had strong decision-making powers and their own locally based officer teams. Middlesbrough District Council has set up eleven community councils which are consulted about any decisions affecting their areas. Glasgow City Council has 104 community councils with membership consisting of people directly elected and people nominated by local organisations. They take part in the council's eight area management committees, advise the council on local issues and resource allocation, and act as pressure groups about issues ranging from health services to rail services. The council funds a community resource centre to provide training, library services, stationery, photocopying and information.

A commitment to decentralised services, and quality in general, demands a corporate approach. This is necessary not only to encourage the development of common values but also to deliver the backup of training programmes, staff resources and information systems which decentralisation needs.

Gaster (1992: 60) describes decentralised service delivery as a way to, 'provide the opportunity for mutual confidence, understanding and knowledge to be built up, both through day to day services and through formal and informal consultation'. A genuine commitment to getting closer to service users can see representatives of community and voluntary organisations appointed to committees and advisory bodies. Tenants' representatives are often co-opted on to area housing committees in local authorities. Users of services can be given a greater individual say, such as York City Council's tenants' choice scheme for housing improvements (described on p. 118). However, account must be taken of people who find it difficult to put forward their views and good support arrangements such as child care have to be in place to encourage people to attend meetings. Research using surveys and focus groups is important to reach people who do not participate in other ways (see Chapter 7).

The area of public policy with most opportunity for the public to comment is land-use planning. Local councils' development plans for their areas have to be available in draft for public consultation before they are adopted and are often the subject of public inquiries, and decisions on planning applications include the opportunity for members of the public to comment. However, people often need assistance in formulating their views and making their case. In planning, this can sometimes be provided by community-based planning or technical aid centres (Forsyth, 1988).

Wider public participation should be a goal across all public services but it needs clear policy and procedure behind it or the result can be public disappointment and cynicism with the process. Gyford cites Manchester City Council's checklist of good practice, which illustrates the kind of issues that can be involved:

1 Both the council and the public need to be clear about the purpose of the exercise from the outset: are the public being informed, consulted or more actively involved in decision-making?
2 It is necessary to identify in advance exactly who is to be involved from among the public: are there specific target groups and do they have particular problems with attending meetings or understanding procedures?
3 A variety of methods needs to be used to reach the public from large meetings to informal groups to opinion polls.
4 Where groups are being regarded as spokespeople for a wider constituency, agreement should be reached at the start on what counts as representation and who is authorized to act in that capacity.
5 Both the council and the public must accept that public involvement will add to the time taken, not only for policy making but also for implementation.
6 Advance agreement should be secured on mechanisms for reporting back to those taking part.
7 The role of council employees must be clarified to all concerned, especially when they are involved in helping groups to formulate their views and to present their case.

(Gyford, 1993: 74)

Decentralisation has an important part to play in the management of quality by improving responsiveness, accessibility and participation. Consultation and participation strategies take this further by being more active in seeking people's views. But there is a growing body of opinion that the best route to quality, in terms of satisfying users and meeting standards, is to adopt a *customer orientation*.

In the private sector, an orientation to customers was based on delivering quality. This was seen as a key factor in the battle for markets in an increasingly competitive environment. Quality assurance (QA) and total quality management (TQM) developed as methods designed to ensure that quality is 'built in' through appropriate systems, working practices and attitudes. In manufacturing industry, quality assurance typically entails the following seven steps:

1 Knowing the customer's needs.
2 Designing a product or service to meet the needs.
3 Guaranteeing the performance of the product.
4 Providing clear instructions for the use of the product.
5 Delivering the product punctually.
6 Providing a back-up service for the product.
7 Using customer feedback to improve the product.

Quality assurance developed in manufacturing industry in response to increasing competition. Sales depend on consistently satisfying the consumer at a price he or she can afford. In public services quality assurance must be justified beyond purely economic reasons. Professional expertise has often been regarded as a guarantee of quality, with public services run by professional officers. But this has been challenged by the rise of consumerism and demands that the consumer of public services should have a right to understand, criticise and choose. There is now also strong political pressure that professionals should be accountable to society as a whole and not just to their professional organisations. This has seen the introduction of published standards and performance indicators. However, quality assurance will not work unless it is 'owned' by the staff who manage and deliver the services. In other words, staff need to be actively involved in planning and implementing measures which assure the quality of their service. Indeed, quality assurance should coincide with an important element of professionalism which is the desire to be self-correcting and self-regulating in providing the best service for the user (Ellis, 1990).

A growing number of local authorities are seeking to meet quality assurance standards for their services. The British Standards Institute's BS 5750 (also an international and European standard) has either been adopted or adapted by many council departments. By the beginning of 1993, BSI Quality Assurance had registered eighty-nine local authorities to BS 5750 for services including leisure, residential social care, cleaning, highways and building (Granville, 1993). Many support services such as finance, administration and IT and computer services are likely to follow.

BS 5750 is based on self-monitoring. This is now widely regarded as the most effective form of quality assurance. Its methods include quality circles which involve groups of staff in regular meetings to discuss how to improve the quality of their service. The more traditional monitoring by specialist inspectors has its place but need not be comprehensive. Random sampling of services with the results reported to relevant managers and committees will generate information encouraging all parts of the service to reduce quality failures. Monitoring, of course, must be complemented by having mechanisms in place to remedy any shortcomings which are identified.

In the private sector, the prime concern of quality assurance is the individual consumer. This remains relevant in public services but there is the additional factor of relevance of the service to the community as a whole. In this respect, there are six dimensions to quality in public services which can be identified. These are:

1 Access to the service, including distance, response times and waiting times.
2 Relevance to need.
3 Effectiveness for the recipient of the service: is it achieving the intended benefit?
4 Equity and fairness.
5 Social acceptability, such as privacy and cultural appropriateness.
6 Efficiency and economy, including workload and unit-cost comparisons.

Some public services are similar to manufactured goods in not involving close engagement with the user, such as refuse collection and technical services. But most, including education, social services, recreation, housing, health and training, involve a personal interaction between the provider and the user. The quality of this experience of interchange is essential to the overall quality of the service. Although there is a danger that customer care may be used as a way of putting a pleasant face on the unpleasant reality that services are underfunded, courtesy, office ambience and speed of service are undoubtedly valued by public service users.

Any policy for quality in public services must be informed about what happens at the point of contact between front-line staff and the public. Because of the level of discretion available to front-line employees in their day-to-day work, quality strategies are unlikely to be successful if staff do not feel committed, involved and supported in providing a good quality service. Gaster (1991) lists the following topics which need to be considered:

1 Do staff have training in interactive skills?
2 Do they have the information needed to answer enquiries from the public?
3 Do they feel that the public have reasonable expectations of what they can deliver?
4 Are they confident enough to deal with verbal or physical violence or threats?
5 Is the working environment demoralising or antagonising to the public?
6 Are staff isolated or do they feel supported?
7 Are they valued and kept informed?
8 What aspects of their work are rewarded?

Most public sector employees are in direct day-to-day contact with the public. Public perceptions of the quality of services are largely based on their experience of front-line staff. The ability of front-line staff to meet the public's expectations depends to a very large extent on how much control they have over the service they are delivering, good planning and training.

Research in this area has shown that inter-personal skills are essential. Front-line staff should be trained in communicating (non-verbal and verbal) and in understanding the 'customer'. Good communication skills frequently have to be brought into play to address difficult or uncomfortable situations. These are by no means always caused by the service failing to measure up to its standards; they can also result from misunderstandings, conflicts with policies or regulations, and unexpected needs.

Training is essential to provide front-line staff with the understanding and skills to put quality strategies into practice. However, staff should be involved continuously in considering the quality of their service. Quality circles are a means of doing this, bringing front-line managers and workers together on a regular basis to discuss the standard of their service. Many methods have been developed in recent years to facilitate feedback from staff. These include staff consultation, staff surveys, staff suggestion schemes and encouraging staff to report problems.

Although front-line employees are key players in quality policies, it is the public's degree of satisfaction with a service which should be used to measure success. The basic measure is whether the experience or perception of a service matches the expectations of it. If services are delivering according to public expectation, then there are also likely to be tangible benefits for staff in terms of their morale and job satisfaction. Low morale is very costly for any employer because it is a cause of time off for sickness and of high staff turnover.

In an evaluation of customer care in a local authority finance department, Blackman and Stephens (1993) used questionnaires to the department's users inviting them to score performance on a series of indicators. The scores could then be used to measure change over time and identify particular aspects of performance which needed particular attention. The following list of indicators is an example of the kind of information that can be sought from users:

1 Do staff perform in a confidential manner?
2 Do staff provide adequate help and guidance?
3 Do staff respond promptly to customer needs?
4 Have staff the necessary resources to do their work?
5 Do staff appreciate users' specific needs?
6 Is a complete service provided?
7 Do staff seek feedback from users?

Blackman and Stephens undertook similar surveys among staff. The following are examples of the indicators they used:

1 Are customer needs given priority?
2 Are customer needs acted upon?

3 Is training adequate?
4 Are staff well-informed?
5 Are there sufficient resources?
6 Are things well organised and goals clearly defined?
7 Is there adequate dialogue with management?
8 Is good performance recognised?

Quality and 'customer care' are overall philosophies and their implementation has to be through a wide variety of specific initiatives. These cannot simply be to survey customer satisfaction and train staff in 'customer care'. In summary, the range of measures involved includes:

1. *Performance standards*. These should be based on the key objectives for the service. Examples include waiting times, amount of service provided, level of use, costs and satisfaction levels. There should be research and consultation to make sure that standards are set with reference to areas that actually concern the public.
2. *Customer input*. This includes surveys, circulating satisfaction cards to users, focused discussion groups and panels of users who are consulted periodically about the service.
3. *Information on services*. Reception staff, leaflets, newsletters and use of the media are the conventional channels for communicating information about services. Special care needs to be taken that information is reaching everyone for whom it is intended. This means using plain English, clear presentation, good information outlets and considering the need for large print and translations. Helplines, exhibitions and talks to the public are other methods used. It is important to evaluate the effectiveness of information, both to test whether the message is getting through and to investigate whether behaviour changes as a result.
4. *Courteous customer service*. This includes induction training for all new staff, programmes of customer care training, and emphasising good practices such as staff identifying themselves personally to members of the public.
5. *Complaints procedures*. These include publicising what to do if things go wrong, response times to complaints, periodic reviews of complaints by management, analysis of trends, and staff briefing on how to handle complaints.
6. *Independent validation of performance*. This includes comparisons with other councils, district audits and performance monitoring. The same point as in (1) applies; performance monitoring should focus on real customer concerns.

Tam (1993) sums up these measures as 'customer management'. They should be part of an organisation's overall strategy. Such strategies, however, can be different in nature. He identifies three approaches:

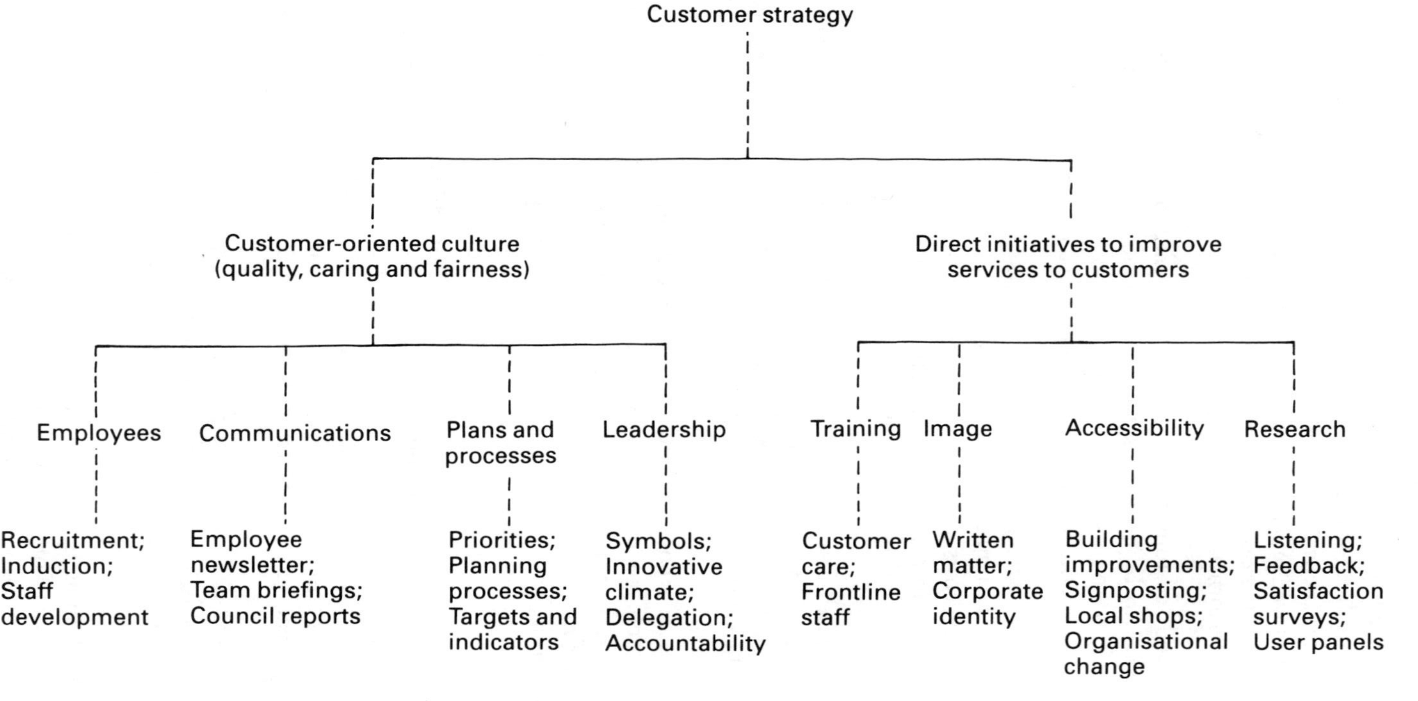

Figure 5.1 A model of customer strategy in local government
Source: Adapted from Hancox, Worrall and Pay, 1989.

1 *The responsive approach.* The central objective of this approach is that users of council services are satisfied with the services they receive. It involves setting up mechanisms to get customer feedback, such as consumer surveys and complaints procedures.
2 *The interactive approach.* This takes customer management a stage further by involving the public in discussions about public services. It should include publicising information which encourages people to judge the performance of services. Mechanisms for implementing the interactive approach include neighbourhood forums and focus groups with users.
3 *The empowerment approach.* This is the highest level of customer management. It gives the public decision-making powers. Examples include tenant management co-operatives and user-controlled projects.

Customer management is a challenge to bureaucratic culture in the public sector. At its best, this culture is one of public service. At its worst, it is one of remote centralised decision-making and blinkered professionalism. Customer management is undoubtedly a culture change for public sector organisations, and therefore has to be introduced in a way that plans change and brings staff along with it.

The nature of an organisation's customer management strategy depends fundamentally on its *raison d'être*: the values and goals which it has and which comprise its 'mission statement'. Defining these values and goals is an essential first step in customer management. This then has to be followed with planning implementation. There are two main aspects to this implementation stage: (a) the introduction of direct initiatives to improve services to customers and (b) measures designed to achieve a customer-oriented organisational culture (see Figure 5.1). It is essential that both aspects are addressed and each one needs to become part of the way the organisation operates.

Individual staff must be clearly linked into customer management strategy. The key to this is *staff development*. Rogers (1990) argues that this can no longer be considered an optional activity. He identifies three main reasons why it is of such fundamental importance:

1 The staff of a local authority are its most valuable resource. What they do is the main determinant of the performance of the authority as a whole.
2 For staff to perform well they need to be clear about what is expected of them, be able to contribute to defining these expectations, and be motivated. They should have the opportunity to review their performance, both by learning from experience and by receiving feedback from others, and should have the support to improve their performance.
3 The organisation should have procedures for planning and appraising the

> work of all staff, ensuring that as far as possible their potential is fulfilled and that they are rewarded.

In practice, staff development usually consists of an interview between the individual member of staff and his or her line manager, normally once a year or once every six months. Both the staff member and the manager prepare for the interview beforehand. This consists of reviewing their perception of the member of staff's job responsibilities, their performance over the past year or six months, and any reasons for performance not meeting targets set at the last interview. These matters are then discussed in the interview and targets set for the forthcoming period.

Staff development is about providing support to help staff achieve their targets and to improve their performance. The actions which can follow from an interview include providing training, removing any organisational barriers or providing opportunities to gain wider experience.

Some public sector organisations link staff development to performance-related pay, although this is at present largely confined to senior managers. Performance-related pay links pay increases to assessments of the performance of individual employees by their manager. It is argued that this provides a financial incentive to improve performance. Indeed, Doern (1993: 25) suggests that service standards, 'may have little staying power if there are not direct links to front-line officials' pay and reward systems'. However, performance pay can be divisive, lead to disgruntlement, and fuel speculation and suspicion among staff. It also rewards good performance but does not provide a way of motivating those staff who are content to perform at a mediocre level. One way of addressing this is to link the pay of all members in a particular section of an organisation to the performance of the section as a whole, so that performance-related pay awards apply equally across the section. This encourages everyone to participate in the success of a team. However, performance pay has to be funded, and in public services with no new money this means either cutting into base pay or cutting services.

Whatever approach is used, it is important to work co-operatively with staff to achieve the goals of the organisation. However no manager can avoid conflict, and when all other means have been exhausted agreed disciplinary procedures have to be used to address those very few staff who may persistently fail to provide a satisfactory service.

THE CONTRACT CULTURE

All staff are having to adapt to a 'contract culture' in public services, whether as commissioners of services or providers of them. The role of contracts in managing public services for quality is (a) to set standards for each service and (b) to establish accountability for meeting and monitoring these standards.

Contracting in the public sector is often a process of negotiation between the provider and the purchaser, in which the views of users are increasingly expected to be represented. Quality assurance is then concerned with ensuring that these agreed standards are met consistently, that actual provision is measured against these standards, and that action is taken to correct any demonstrated weaknesses.

Contracting thus makes defining service levels and standards an explicit process. The bureaucratic model of public services has been one where the same organisation assesses need for a service and provides it. Contracting introduces a purchaser–provider split. This separates needs assessment and contracting/commissioning functions from operational matters, which become the concern of provider organisations. In separating the interests of the provider/contractor from those of the client there is greater emphasis on value for money. This will not just be a concern with cost but also with the relevance and quality of the service purchased by a client department. The relevance of a service is the extent to which it meets assessed needs. Its quality is usually assessed in terms of user satisfaction.

There are two types of contract that have arisen specifically in response to quality concerns: customer contracts and internal contracts.

Customer contracts are guarantees to the users of public services. They describe the service and specify its cost, the standards against which it can be measured and how to complain if the service fails to meet these standards. Customer contracts are usually distributed as leaflets to service users. The following is an example from the Northern Ireland Housing Executive, Northern Ireland's public sector housing authority (NIHE, 1992). In this extract from its leaflet, the standards of service that someone applying for a home can expect are explained:

> We will provide you with an application form and advise you how to complete the form. We will also advise you about areas in which housing is immediately available and about other landlords who may be able to help you.
>
> **Standards**
>
> (a) We will acknowledge your application within 3 working days.
> (b) We will visit you within 2 weeks.
> (c) Within a further week we will place you on the waiting list.
> (d) We will give you the opportunity to view any property we offer you before you decide whether or not to accept the tenancy.
> (e) We will give you access to the information you provided in your application. You can request it at any time before or after becoming a tenant.

The leaflet includes a card which can be returned if it is felt that the Housing Executive falls short of its service standards.

Customer contracts often include a section such as, 'what we ask you to do'. This reflects what Doern (1993) terms the 'co-determination' of public services – the fact that their performance depends on their users as well as on the providers. For example, the Northern Ireland Housing Executive states that:

> When you apply to the Housing Executive for a home you have the *right* to a free summary of our rules for deciding the order in which we allocate houses; to study the full rules at your District Office; to buy a copy of the full version.
>
> You are *responsible* for providing full and accurate information about where you live and about yourself and your family; informing us if there are any changes in those circumstances.

York City Council, the first council in the country to introduce customer contracts, has extended the concept to people applying for jobs with the council. Guarantees are given regarding realistic timetables for applications, keeping in contact while the application is processed, fair selection procedures, an easy-to-use complaints procedure and training to help new employees. The council feels that how job applicants are treated shapes how the council is viewed locally and nationally (*Municipal Journal*, 5–11 November 1993).

Internal contracts or *service level agreements* (SLAs) are now widespread in local government. An SLA describes the service which a central department such as finance, legal services or computing services will provide to a direct service department such as housing or social services within the same authority. It states the quantity of the service that will be provided (usually over a year) and the cost that will be charged. At least a proportion of this is negotiable and the service department is essentially a customer that buys what they need. This has ended the practice of charging out central services as overheads on direct service departments.

From 1995/6, the central corporate and construction-related services of local authorities are required to publish a 'Statement of Support Service Costs' (SSSC) as part of their annual accounts. Compulsory competitive tendering is being phased in for these services, starting with 45 per cent of an authority's legal work from April 1996, but the concept behind SSSCs is to encourage the development of an internal market. The government's objectives for SSSCs are that they should demonstrate the cost to an authority of providing support services, whether in-house or out-sourced, and stimulate the challenge of these costs by internal customers.

Hoggett (1991) contrasts this new contract culture with the bureaucratic model of large public sector organisations directly providing services. Control in bureaucracies is through centralised hierarchies. In the post-bureaucratic organisation, control is by contract. While bureaucratic control by hierarchy placed emphasis on methods and procedures within the organisation, control by contract puts the emphasis on results. Results may be seen very much in terms of containing costs and controlling expenditure. More recently, however, quality from the user's perspective has been highlighted. Important consequences follow from this:

> Formalising the relationship between the local authority and its consumers through standards and contracts of service, while empowering consumers to make informed judgements, has implications for performance measurement. Local authorities will not only have to produce and use internal data but develop ways of linking this assessment with the consumer's judgement.
>
> (Local Government Management Board, 1992b: 12)

A recent TUC report on *The Quality Challenge* emphasises that it is not necessarily the case that an older person will regard five frozen meals a week as a substitute for a daily meals-on-wheels service, even though such a change may increase performance (TUC, 1992). Similarly, improving the quality of provision by making more coherent use of existing resources is obviously desirable from a consumer point of view, but consumers often want an increase in the quantity of provision which can only be achieved by the allocation of substantial and genuinely new resources to the service.

The contract culture clearly has advantages in introducing clarity into decisions about levels and standards of public services. But it has also caused problems, especially when significant detail in service specifications is necessary and when there are a large number of providers to monitor. As Chapter 3 discussed, this can add to costs, possibly cancelling out any value for money gains. In addition, in a competitive situation contracts can force providers to reduce costs by cutting the pay and conditions of workforces (see Chapter 2).

CITIZENS' CHARTERS

Citizens' charters are a way of making an explicit public commitment to 'customer rights' in public services. York City Council led the field with the first of its annual citizens' charters in 1989, committing the authority to performance monitoring and open government. The Labour Party's national policy review saw the launch of its citizen's charter in 1991, followed soon afterwards by that of the Conservative Party. The Liberal Democrats' version is the 'people's charter'.

Certain values are common to all of these charters, such as quality, regulation, published performance indicators, publicised complaints procedures, compensation and institutional change. There are, however, important differences. The Labour Party opposes compulsory competitive tendering and advocates customer contracts and the training of staff to achieve quality standards. It has proposed replacing the government-appointed Audit

Commission with a Quality Commission, placing the auditing of service quality on an equal footing with financial audits of efficiency and effectiveness. The Liberal Democrats' people's charter includes commitments to decentralisation, the introduction of consumer panels to monitor performance and complaints, and a greater use of consumer surveys. The Conservative Party's approach has been dominated by extending competition. The two main strands of this are compulsory competitive tendering for manual and professional services and the use of published league tables of performance for services such as schools and housing management.

York City Council's first charter was delivered to every household in the city in April 1989. It was both a mission statement and a contract which stated targets for services. A second charter was published the following year, simplified and made more concise as a result of feedback from the first charter. As well as stating targets, it reported back on how well the previous year's objectives had been met and the reasons for instances where these had not been met. The 1994/5 charter covers the following topics: jobs, homes, equal opportunities, the environment, cleansing and health, transport, leisure, customer care, council management and local democracy. For each topic the charter gives a brief background to the issues, sets out the council's policy, states its aims, lists specific commitments and gives a contact for further information. The 1994/5 charter also includes a report back on 1993/4, detailing what the council did not achieve and why. The charter is a distributed to all households in a newspaper format. Table 5.1 shows as an example the equal opportunities section of the charter.

York has also been at the forefront of giving public service users more choice. It pioneered 'tenant's choice' whereby council tenants in modernisation schemes can choose what is done to their home and what fittings and materials are used. Tenants choose the contractor, and the contractor works for the tenant, giving maximum control over the work being carried out in their home. The quality of work is checked after completion and in addition tenants receive customer comment cards and can make use of the housing department's general complaints mechanism if they are dissatisfied. Surveys have demonstrated very high levels of satisfaction (York City Council Housing Services, 1993a).

Public opinion surveys have found most people in favour of charters, although common criticisms are that they are too vague and do not use straightforward language (Gosschalk and Page, 1993). The public are interested in specific commitments and information, such as that letters will be answered within five working days and the telephone numbers and opening hours of individual services. Some of these types of indicator are required to be published by the national citizen's charter and many councils now include specific information of this nature in 'customer contracts' which are widely distributed to residents. There is evidence of a strong correlation between how well-informed people feel and how highly they regard services.

Table 5.1 An extract from *The York Citizen's Charter 1994/5*: 'Opportunities for Everyone'

There is overwhelming support among York citizens for spending on services and facilities for people with a disability.

We want every York citizen to have the same opportunities, and the standards of service and quality of life they need. We make sure our services and decisions reflect the needs of all York people.

This year we will:

* make our largest ever single investment – £100,000 – in providing 230 dropped kerb crossings for people with disabilities and parents with pushchairs;
* improve access for people with disabilities to our Housing Services Department at 4 St Leonard's Place;
* provide special warm water swimming sessions for people with disabilities;
* train some Reception staff in sign language;
* promote the Council's child-friendly York Award Scheme, encouraging businesses to provide child-friendly services and facilities;
* adapt over 600 homes for disabled and elderly people.

Further information: Citizen's Unit, York 613161, ext. 1025.

Source: York City Council, 1994.

The Conservative government's national citizen's charter was launched in 1991. It has the status of a White Paper, although it is not being legislated as a whole. Some of its recommendations are to persuade, and others are being carried out in new laws. It applies to all public services, including central government departments and agencies, nationalised industries, local government, the NHS, the courts, the police and the emergency services. Most public services will publish their own charters and all organisations will publish their service standards. The charter's main elements are:

1 Local councils must set standards for their services, publish these standards and provide remedies if standards are not met. The Audit Commission has the legal power to publish league tables showing how different councils compare.
2 All parents must receive a written report on their child from their child's school at least once a year. School-by-school examination results and summary reports of school inspections carried out every four years must be published. Plans to publish league tables of National Curriculum test results for 7 and 14 year olds were abandoned in the face of criticisms that they would be unreliable, although league tables for tests of 11 year olds are due to be published after 1995.
3 Council tenants have an improved 'right to repair' and a new 'right to improve' by which they can receive compensation for home improvements they have carried out if they move house. An Ombudsman

scheme is being introduced to deal with housing association tenants' complaints.

4 In the National Health Service, a guarantee has been introduced that no patient will wait more than two years for admission to hospital. Since April 1993, a waiting time of eighteen months has been guaranteed for hips, knees and cataracts treatments. If an outpatient has to wait more than thirty minutes, an explanation must be provided. Family doctors must display information about their services in waiting areas and leaflets.
5 Measures have been introduced to speed up and co-ordinate road works; to improve the complaints procedures of the gas, water, telecommunications and electricity companies; and to strengthen schools and social services inspectorates.

A few simple principles are behind these changes:

1 Setting, monitoring and publishing explicit standards for the services that individual users can reasonably expect to receive. Publication of actual performance against these standards.
2 Full, accurate information readily available in plain language about how public services are run, what they cost, how well they perform and who is in charge.
3 The provision of choice wherever practicable. Regular and systematic consultation with service users. Users' views about services, and their priorities for improving them, should be taken into account in final decisions on standards.
4 Courteous and helpful service from public employees who will normally wear name badges. Services available equally to all who are entitled to them and run to suit their convenience.
5 An apology, full explanation and a swift and effective remedy if things go wrong. Well-publicised and easy-to-use complaints procedures with independent review wherever possible.
6 Efficient and economical delivery of public services within the resources provided. The independent validation of performance against standards.

The Local Government Act 1992 requires the Audit Commission to publish nationally information about local councils' performance as part of the citizen's charter initiative. The first of these national league tables will appear early in 1995. Some 150 indicators will measure the performance of local government services and, in many cases, enable comparisons to be made across councils. The standards themselves are set by the elected members of councils, in consultation with service users, except where the standards are laid down in legislation. For example, councils will have to state repair times for different categories of repair reported by council tenants, but they may adopt different times and categories according to local circumstances.

However, the standard that all parents receive a report once a year on their children's progress at school is a legal right established by the Education (Schools) Act 1992.

In December 1992, the Audit Commission issued its first list of performance indicators which it required local authorities to collect and publish for 1993/4. These are intended to be benchmarks to help the public judge the effectiveness, efficiency and cost of services provided by every local council. The selection of indicators followed consultation with local authorities, consumer organisations and government departments.

The list of indicators has three main aims. First, to enable people to judge the performance of their councils in providing key services and how this changes over time. Second, to show how well councils are performing in comparison with similar authorities. Third, to enable local authorities to explain to the public their own targets and how they are doing against them. Local councils must publish their performance in a local newspaper.

The Audit Commission revises the indicators annually but avoids substantial alterations to achieve consistency from one year to the next. For example, for 1994/5 it added a number of new indicators for leisure services, highways, street cleaning, environmental health, trading standards and the care of children. It also added some indicators designed to relate more clearly to quality. As an example, Table 5.2 lists the indicators which relate to council housing services.

This move towards explicit accountability on the basis of measurable indicators has been heralded as a way of empowering citizens with information on their local councils. It is intended that comparisons will be possible both within authorities from year to year and across authorities.

The citizen's charter initiative is having a positive effect in stimulating local councils to inform the public about their services and think carefully about the business they are in. Local government exists to serve the interests of its local population. Information about how it does this is essential to enable people to make demands on, and informed judgements about, local services. However, the initiative is very prescriptive and takes little account of local priorities, as reflected in the following introduction to York City Council's own 1993/4 charter:

> Next year we will be measuring our performance against standards set by the Government. We believe that it's *your* right to say what our standards should be. They should be based on *local* concerns. So, in the year to come we'll ask you what those targets should be, and let you know how well we've done against those, as well as against Government standards.

The charter is, 'a national template, rather than something to be negotiated locally in response to service conditions and public opinion' (Local Government Management Board, 1992b: 3). Proponents of national indicators argue

Table 5.2 Citizen's charter performance indicators 1994/5: housing

Citizens' questions (the issues covered by the indicators are listed below in the form of questions which citizens might ask)	*Indicators*
The provision of housing accommodation	
How many council homes are there, and how many have been adapted for the elderly or disabled?	1 The number of dwellings managed by the authority at 31 March 1995. 2 The percentage of these dwellings adapted for elderly or disabled people.
How many flats have controlled entry?	1 The number of flats in blocks of 3 storeys or over managed by the authority at 31 March 1995. 2 The percentage of these flats with controlled entry.
How many houses have been let to tenants on the waiting list?	The number of dwellings let to new tenants: 1 Authority dwellings; 2 Authority nominations to housing associations; and the percentage of these let to: 3 Homeless households. 4 Others.
Housing the homeless	
How many homeless households are in temporary accommodation?	The number of homeless households in temporary accommodation on 31 March 1995 and housed in: 1 Bed and breakfast accommodation. 2 Hostel accommodation. 3 Other accommodation.
How long do they stay in bed and breakfast accommodation or hostels?	The average length of stay in bed and breakfast and hostel accommodation.
Re-letting and repairs	
How quickly are empty homes re-let?	The percentage of dwellings that are empty: 1 Available for letting or awaiting minor repairs. 2 Others. The average time taken to re-let dwellings available for letting or awaiting minor repairs.
How quickly are repairs carried out?	1 The number of repairs requested by tenants at each priority level set by the authority.

Table 5.2 continued

	2 The authority's target response time(s) for each priority level. 3 The percentage of jobs completed within target time(s).
Are appointments offered for repairs, and are they kept?	1 The authority's policy on offering tenants appointments for repairs to be carried out. 2 The level of performance: (a) the percentage of repair jobs for which an appointment was offered; (b) the percentage of repair jobs for which an appointment was made; (c) the percentage of appointments that were kept by the authority.
Rents and costs	
How successful is the council at collecting rent?	The rent collected as a percentage of the rent due.
How many tenants are significantly behind with their rent?	The percentage of all tenants owing over 13 weeks' rent at 31 March 1995, excluding those owing less than £250.
What is the average rent for a council home?	The average weekly rent per dwelling.
What is the rent spent on?	The average weekly costs per dwelling, itemised as follows: (a) management; (b) repairs; (c) bad debts; (d) empty properties; (e) rent rebates; (f) capital charges; (g) other items, net; (h) less government subsidy. (i) Total = average rent.
How much is spent on improving council homes?	Capital expenditure per dwelling on major repairs and improvements.
The payment of housing benefit and council tax benefit	
How quickly does the council pay benefits?	1 The number of new claims for council tax benefit. 2 The percentage of such claims processed within 14 days. 1 The number of new claims for housing benefit from local authority tenants.

Table 5.2 continued

	2 The percentage of such claims processed within 14 days.
	1 The number of successful new claims for rent allowance.
	2 The percentage of such claims paid within 14 days.
How much does it cost the council to pay benefits?	1 The total number of benefit claimants.
	2 The gross cost of administration per claimant.

Source: Audit Commission, 1993b.

that local indicators do not enable councils to be compared because they will vary from authority to authority. However, crude national comparisons will not compare like with like. In order to compare standards across services or authorities, their capability should be broadly equivalent. This is not the case because the standard spending assessments (SSAs) used to achieve need and resource equalisation between authorities are too crude as means of providing authorities with equivalent financial capability to achieve pre-defined standards in each individual service (Audit Commission, 1983a; see also Chapter 4). Although SSAs may be acceptable in principle as a means of allocating central government grant as a block to local authorities, they do not place all local authorities in the same financial position for every service.

The comparison of performance has been particularly controversial in relation to the publication of school-by-school examination results. Because examination achievement is mainly a function of the prior attainment of the pupil, comparisons between secondary schools will largely measure differences in the levels of performance of the pupils at intake rather than the effect of the secondary school. This latter 'school effect' is known as *value-added* (see Chapter 8). It is a measure of the enhancement of pupils' learning over time achieved by the school, and is quite different from crude examination results, which have no value in making school-by-school comparisons because they do not compare like with like.

There has been an increasing amount of central control over local government spending in recent years but local authorities do still adopt priorities which influence service levels and standards locally. The national citizen's charter and the Audit Commission's indicators have very little regard for such local policy priorities, as Rodrigues comments:

> Unless there is complete agreement between the policy priorities implicit in the commission's measures and those of each local authority the new indicators will in effect cut across the latter. This will have a

distorting effect on local priorities and local rationing decisions, skewing organisational processes to a centrally determined perspective. Messages to staff will be confusing. Just whose priorities are they supposed to be working to?

(Rodrigues, 1992: 13)

THE EFFECTIVENESS OF POLICIES

Whilst the 1980s were primarily concerned with efficiency, the 1990s have seen more emphasis on effectiveness. This has had an influence on both services and policy-making. Evaluating the performance of policies is as important as evaluating the performance of services. 'Policy performance' is a question of monitoring and reviewing the effectiveness of policies. Two fundamental questions are involved: 'what is the policy trying to achieve?' and, 'are the results of the policy worth while'?

There is often resistance among policy-makers to monitoring the results of their actions. Reade suggests the following reason for this:

> Few public policies achieve exactly the outcomes envisaged by those who urged them, and many produce results which are positively embarrassing, often because the context in which they are operated is complex and changing, and thus affects them in ways which cannot be foreseen. It is always more pleasant, therefore, for policy-makers and their expert advisers to concentrate on devising new policies, rather than on studying the consequences of previous or existing ones.
>
> (Reade, 1987: 70)

However, as Reade states, 'intelligent policy-making is literally impossible unless it is based on knowledge of its own past consequences' (Reade, 1987: 70).

A policy is essentially a frame of reference for decision-making. The starting point for policy-making is a pre-existing state of affairs which is considered unsatisfactory and which a policy attempts to change. The mechanisms for achieving change are *policy instruments* such as personnel, plans, expenditure or regulations. These are used to change the existing state of affairs to a state that is judged more desirable.

Policy-making must thus start with clear objectives. Policy seeks to achieve these desired objectives through its effects on the decisions made by the people the policy addresses. The constellation of people addressed by a policy, and therefore with an interest in it either as implementers or as intended beneficiaries, constitute a 'policy network'. The degree of coincidence between planned decisions to realise a policy and actual decisions in practice will reflect the degree of control available, especially the impact of changes in external circumstances.

Some examples of policy objectives are to ensure that there is enough land available to meet present and future housing needs, to create employment and training initiatives that are taken up by people in long-term unemployment, to stimulate and support community groups, to reduce deaths and injuries from accidents, or to limit traffic growth. Actions are then planned to achieve these objectives.

Objectives are often defined too loosely, and are then essentially general indications of direction rather than operational goals. Although it can be important to identify overall aims at a more general level, objectives should be specific and relate to particular policy areas. They should reflect and distinguish between different interests. For example, in land-use planning policy these interests could include the possibly conflicting ones of economic growth, quality of life and the long-term protection of resources. Whenever possible, the priority to be given to particular objectives should be stated.

Targets should also be specified for each objective. These are then used as measures of achievement. There can be long time lags between policy formulation and implementation, and changing circumstances may outdate the policy before it is implemented, causing deviations from the initial policy. For this reason, 'milestones' are often used to track progress. Milestones are pre-defined indicators of what intermediate outputs are expected, and when. This enables problems with delivering the final output to be diagnosed at an early stage; either corrective action can then be taken or the policy can be revised and new milestones established.

Policies must be evaluated over the appropriate time-scale; the evaluation of land-use planning policies, for example, could cover the typical forward planning period of up to 15 years.

In summary, policy-making and implementation have the following stages:

1 *Identification of objectives.* What is the policy trying to achieve? It is useful to have a benchmark picture against which change can be measured and to specify targets which represent achievement of the objectives.
2 *Policy inputs.* These are the policy instruments by which the objectives are to be pursued and reflected in decisions. They will include personnel, budgets, plans, services, grants, guidance, regulation, etc. The effectiveness of policy instruments is an important question in itself.
3 *Intermediate outputs.* These are the effects of the policy in practice. They can be good or bad, expected or unexpected. The expected effects are those identified in the objectives for the policy. The effect of banning through traffic from a neighbourhood may be as expected in terms of quietness and seclusion for its residents. But an unexpected effect may be to subject people living nearby to much increased levels of noise and danger as a result of the displacement of traffic from the other

neighbourhood. Neighbourhood improvement programmes have also been shown to have unintended effects in displacing the poorest sections of the local population elsewhere. Theoretically, the evaluation of effects requires examination of what would have happened in the absence of the policy. This is sometimes possible when a policy is introduced on a pilot basis in one area but not another. But generally evaluation has to be on the basis of reviewing trends and making specific case studies which examine how individual decisions have been influenced by the policy. Approximate knowledge is often all that is possible. Much can be achieved, however, with monitoring systems that at least identify whether expected changes are occurring.

4 *Impacts*. These are the consequences of the policy. A policy will have a range of impacts which can be represented as its costs and benefits, both monetary and non-monetary. Not all these impacts will be perceived. For example, the consequences of planning are often evaluated in terms of the property market, such as the effect of planning controls on house prices. Prices may rise because of increased demand caused by planners refusing to allow greenfield land to be developed, thereby causing new entrants to the housing market to bear some of the 'costs' of planning. A less common question asked about planning is its consequences for social polarization (see below).

5 *Evaluation of the achievements of the policy*. This entails analysing how far the policy has achieved what was intended and how far the results have been worth while. It is concerned with both the effects and the consequences of the policy.

Table 5.3 shows how this model can be used to evaluate a local authority's policy aimed at protecting the viability of town and local shopping centres by using land-use planning controls to oppose new out-of-town shopping centres. In this example, the policy has failed because a developer won an appeal to the Department of the Environment against the council's decision to refuse planning permission for an out-of-town development. The consequences of this are identified using headings based on the four key objective areas for planning policy. Such assessments will contain a substantial element of judgement but this can be reduced by group discussion and agreement.

Rees (1988) researched the consequences for social polarisation of a similar example in Swansea. She showed how the closure of a Tesco store in a district centre led to a marked polarisation between low-income carless households, who had no choice but to continue shopping in the district centre despite a marked reduction in shopping opportunities, and car-owning households, who were able to shop in a new Tesco superstore opened in Swansea's enterprise zone. Rees found that most car-owning shoppers would have preferred to continue shopping in the district centre, a choice denied by the closure. She considers the policy implications and concludes that planning

Table 5.3 Evaluating a planning policy

Objective

- to maintain and enhance the role of existing town and local shopping centres.

Policy target

- opposition to edge-of-town shopping.

Achievement

- low – major Safebury's out-of-town proposal granted on appeal.

Consequences

Property market:	Trade diversion resulting in low levels of retail growth and investment in Anytown Centre.
Economic:	Estimated 5 per cent growth in retail employment as a result of attraction of trade from surrounding centres. Loss of three local food retail businesses in Anytown Centre.
Quality of life:	Residents enjoy improved access to competitive shopping facilities.
Environmental:	Loss of five hectares of grade 3 agricultural land and erosion of open countryside between Anytown and Someville.

Source: PIEDA plc, 1992.

can have a role in steering large stores to established district centres. However, the subsequent closure of such monopolistic stores may inflict an even greater blow on district centres than locating elsewhere in the first place. Linking large stores to frequent and convenient public transport might be a better policy response.

Figure 5.2 shows an example of an 'adapted balance sheet' used to identify the impact of individual policies or instruments in land-use planning. The objectives for this policy area are grouped under economic growth and development; quality of life; and environmental conservation. Quality of life objectives could include ensuring the provision of shopping facilities to meet the needs of the population; preventing noise, smells or excessive traffic generation; maintaining health and safety; and facilitating enjoyment of the countryside and leisure time. Examples of possible measures of achievement for this objective are level of, and accessibility to, retail provision; number of complaints about noise and nuisance; road accident statistics; and level of, and accessibility to, open space, countryside, sport and leisure facilities. The aim is to develop a set of indicators which can show how aspects of quality of life change over time. The consequences of policies are likely to be both intended and unintended. For instance, improved retail provision might be offset by a deterioration in accessibility arising from increased traffic congestion.

The adapted balance sheet approach can be applied to other policy areas if objectives, inputs, intermediate outputs and final outputs are specified. It seeks to take account of all the advantages and disadvantages of policy decisions (PIEDA plc, 1992). Reade (1987) suggests that planning balance

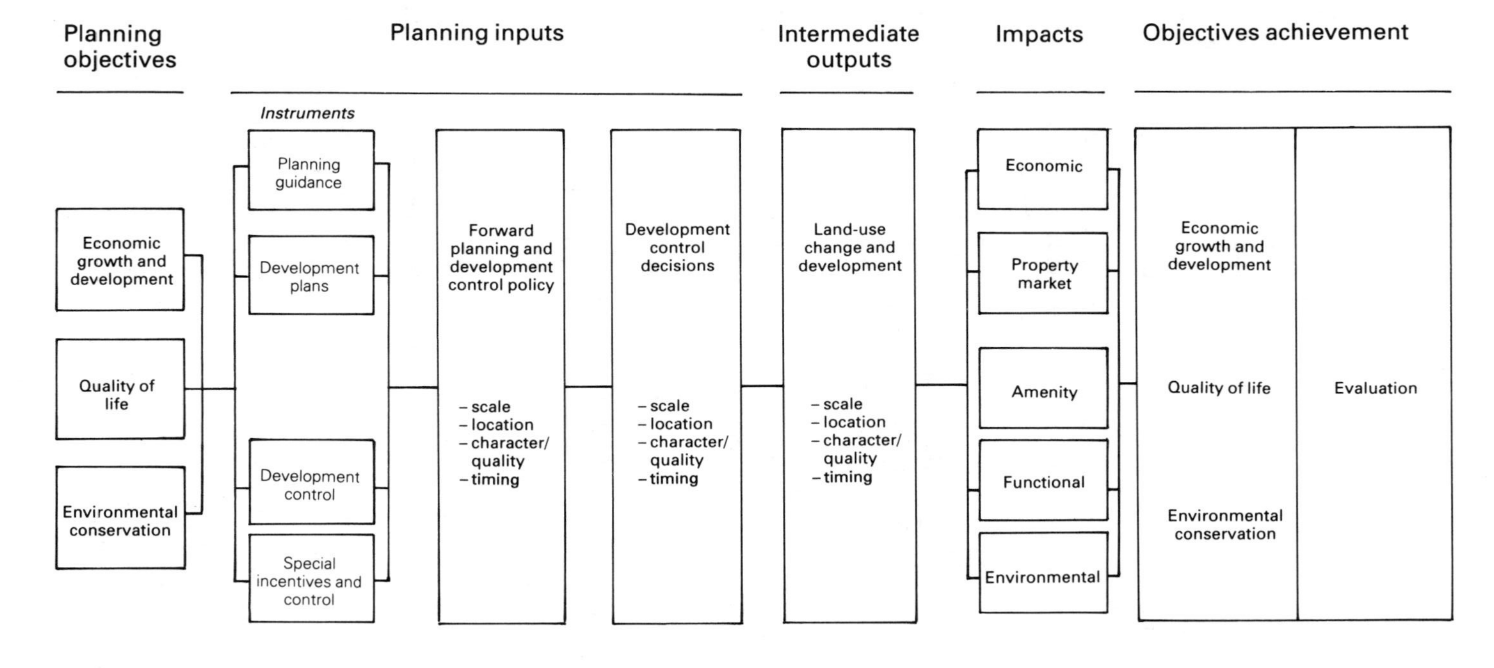

Figure 5.2 The adapted balance sheet
Source: PIEDA plc, 1992.

sheets could be used to explain to the public the expected economic and social effects of policies and to monitor the extent to which the impacts are anticipated accurately.

Many policies seek to achieve a balance between more than one objective. Land-use planning has to balance the objectives of economic growth and development, protection and enhancement of the quality of life for the present generation, and conserving environmental resources for future generations. Indeed, Healey (1982) describes land-use planning as the management of complex processes of change through mediation between interested groups with conflicting interests and powers. This is equally true of other policy areas.

Policy implementation is usually not straightforward. Policies are often directed at people who have their own goals and their own means to achieve these goals. They may also interpret policies differently. People do not mechanically implement policy along the lines set by policy-makers, and extensive negotiation or compromise may be necessary. Policy outcomes usually depend on changing the behaviour, status or attitudes of individuals, households or companies. This means that co-operation is needed, with the policy-maker having greater or lesser power to bring about this co-operation depending on the instruments available. In many ways, the effectiveness of a policy is reflected in how much use people addressed by the policy make of it in their decisions. This in turn depends on how useful the policy is to tackling the problems that people face.

BUSINESS PLANNING

Business planning is the method by which policies are put into practice. It is now widely accepted that public services should be 'needs led' and not led by the interests of services that are already being provided. It should not be assumed that services are well-targeted on needs. People's needs change and neither policy-makers nor managers can assume they know what these needs are. There are many ways that needs can be assessed, and two key approaches using research and community development are described in Chapters 6 and 7.

Once needs are assessed the results have to be linked to how decisions are made. As Tam comments about local authority business planning:

> Customer needs, identified objectively without distortions, are what should determine the fundamental direction of local authority management. If this is to happen in practice, the needs as perceived must be connected through some steering mechanism to the engine of the management process. This steering mechanism is available in the form of target setting procedures.
>
> (Tam, 1993: 4)

Business plans link objectives for services, derived from assessments of need, to measurable targets and indicators of performance. A business plan will include a clear statement of the overall strategy of the organisation, whether a purchaser of services or a provider. It often states the actions the organisation will take over a given time period to manage both threats and opportunities.

Business planning is now regarded as good management practice in the public services. This has coincided with government legislation during the 1980s and 1990s which is privatising public services and introducing market forces into decision-making. Business planning is often criticised as a further example of this strategy, but this is an inaccurate interpretation. Business planning has become much more important with the introduction of compulsory competitive tendering, the client–contractor split, performance standards, league tables, opt-outs and privatisation. However, there are also overriding democratic justifications for its implementation. It is a means of setting targets based on open and thorough discussions of what people actually need, rather than pre-conceived ideas and personal interpretations of 'what is best for people'.

Delderfield, Puffitt and Watts (1991) distinguish between *primary objectives* and *secondary objectives* for public services. Primary objectives are derived from the mandatory statutory functions which central government in Britain allocates to local authorities and other public bodies. The actual means of achieving these primary objectives can vary markedly according to local policy. Delderfield, Puffitt and Watts describe these means as secondary objectives. Many local authorities will attach primary importance to functions which are discretionary rather than mandatory because there is a particular local political commitment to the objective. Examples include equal opportunities or tourism. However, at a time when local authorities are being forced to bring spending into line with central government assessments of the need to spend, separating the mandatory functions that have to be carried out from the discretionary functions that may be causes of 'overspending' has become an essential part of public sector management (see Chapter 4).

A business plan for a public service can be organised into three broad areas:

1 *Impact*: statements of policy objectives for the service's target groups. This involves answering: who is expected to benefit?; how are they expected to benefit?; how will this benefit be measured? This last question is about the service's fitness for purpose – does it do what it is supposed to do? The starting point is to draw up an inventory of what the organisation does, the needs which each activity is meeting, and any evidence about unmet needs.
2 *Service*: statements of policy objectives for key *processes* or activities.

These processes are designed to deliver the policy objective for the target group, e.g. care management in social services or development control in land-use planning.

3 *Logistics*: statements of policy for the generation and use of *management information* about both impacts and processes. This involves defining the data needed to construct performance indicators. This is likely to include a process of regular reporting of indicators from administrative records, as well as specific performance reviews which investigate an operational area in more depth. The first requires a database from which information can be extracted and aggregated. The second requires a research approach, using surveys or focused discussion groups with users and staff.

Thus, the business plan identifies what each service is meant to deliver. The actual delivery of the service without defect, on time and within budget, is another process which has to be managed.

Objectives should be translated into numerical *targets* whenever possible in order to facilitate the measurement of performance against these targets. Qualitative indicators have a role if they can be reported in a way that monitors how well the organisation is functioning. However, qualitative aspects are often best investigated using specific studies of topics such as the way resources are used, the way decision-making criteria are applied or the views of service users.

Targets should not be imposed by a central bureaucracy, based on what decision-makers think are suitable targets for users. They should be user-led and derived from close interaction with the local community. Target setting should involve a process with the following four stages:

1 Involve users in the development of targets.
2 Specify input and output targets. These can either be incorporated into a charter document or used in 'customer contracts' for specific services. A customer contract will describe the service, set out its targets, such as frequency, times and standards, and explain how to complain. This is usually widely distributed as a leaflet. Output targets should always be traced back to establish the input targets which are necessary, checking that resources are sufficient to meet the targets identified.
3 Report on progress. This often takes the form of an annual report.
4 Explain shortfalls.

Targets should be:

1 Meaningful and interesting to members of the public.
2 Measurable and auditable.
3 Describe the service adequately and must not encourage distortions in the balance of services.

4 Easy to interpret and allow for comparisons over time.
5 Cheap to collect, publish and audit.

The indicators used to measure performance against targets should not be so many that they are unmanageable, nor so few that significant variations cannot be signalled. Jackson and Palmer (1992) suggest that no more than six *key indicators* should be used for a specific activity. 'Indicator' is a better term than 'measure' because direct measurement of efficiency or effectiveness is very difficult. This is because there rarely exists an unambiguous causal relationship between the decisions managers make and the efficiency or effectiveness of the service.

Care needs to be taken in constructing and interpreting indicators. Per capita cost measures, for example, could be based on the total population in a local area, the total number of users or the total number of potential users. If total population is used and this is falling in one authority and rising in another, and each spends an identical amount on a service, then in the first authority per capita expenditure will be rising, whilst in the second it will be falling. This does not mean that the first is less efficient than the second. Ratios are often used, such as the number of users of a service as a percentage of the total population. A change in a ratio, however, can be caused by a change in the numerator, a change in the denominator, or both.

It is important not to confuse indicators of activity with indicators of performance. Activity is an intermediate stage and means to an end. Users of services will judge service quality in terms of both intermediate outputs which they experience, such as the quality of reception services and the time they wait, and the final output, such as satisfaction with a housing repair. This is illustrated in Figure 5.3. Intermediate measures relate to the amount and quality of provision, such as units of housing with care and support for people with mental health problems, or training and education places for 16–24-year-olds. These measures enable programmes to be tracked for whether they are on course. Final output measures relate to level of utilisation of services, satisfaction with them, and impact on key objectives such as an increase in the number of people with mental health problems living in ordinary housing or an increase in the number of young people with training and education to NVQ level 3. As Figure 5.3 indicates, economy and efficiency are also issues in public service delivery and local councils have to balance an emphasis on effectiveness with an emphasis on efficiency.

Quantitative performance indicators will point to where problems are occurring; they are signals and usually only suggest where things need further investigation. More in-depth *performance reviews* are usually necessary to find out why actual levels of achievement are not meeting expectations. This might include a review of the messages which the service surroundings give to both users and staff. For example, what are staff and user views of the physical appearance and comfort of public offices and

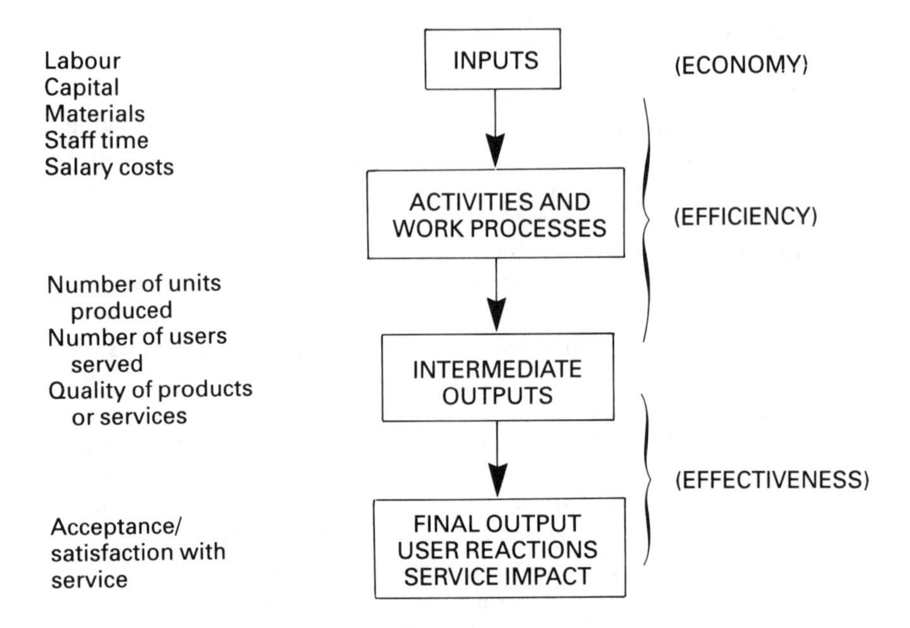

Figure 5.3 Stages in the delivery of public services
Source: Jackson and Palmer, 1992.

reception areas, queuing systems, signposting, and so on? Thus, quantitative indicators should be combined with qualitative performance reviews.

Warwickshire County Council has developed performance indicators for social services to young people. Table 5.4 shows how the three broad areas of a business plan – impact, service and logistics – are translated into objectives and performance indicators for diverting young offenders from custody. The 1989 Children Act exhorts local authorities to take steps to prevent young people from committing crime. Juvenile crime frequently hits the headlines and although persistent offending has been increasing, the number of young adult offenders has in fact declined in recent years. The decline in the number of offenders is partly due to the decline in 10–17-year-olds in the population, but it is also linked to a reduction in custodial sentencing and the use of alternatives to custody. These include counselling support, supervision orders, community service and probation orders, bail accommodation and family placement.

In introducing performance indicators, it is wise to pilot them first and to acknowledge that they will need refinement over time. The development phase should include training key staff and involving them in the design of the indicators. Their use within an organisation also needs to be determined

Table 5.4 Warwickshire County Council: performance indicators for a policy of diverting young people from custody

Objective	*Performance indicators*
Impact objective	
To divert juveniles from custody	Proportion of juveniles receiving custody within a period
	Comparison within community
	Comparison with national data
	Offending characteristics of those receiving custody
	Previous offence career of those receiving custody
Service objective	
To provide appropriate community-based resources as alternatives to custody	Proportion of juveniles receiving community service/supervision orders
	Offending characteristics of those receiving community service/supervision orders
	Previous offence careers of those receiving community service/supervision orders
	Comparable data on juveniles receiving custody
Logistical objectives	
To recruit and train specialist staff to run alternatives to custody	Target date to conclude recruitment
	Analysis of recruits by person specification
	Analysis of recruits by job specification

Source: Social Information Systems, 1991.

within a strategic framework, with reporting to managers, providers and a performance review committee.

Business planning is much more than stating targets. The core of the activity is to analyse the factors which might affect the department's ability to reach its targets. It is common for this to be a SWOT analysis. This is usually done in a group and involves identifying:

1 *S*trengths, *W*eaknesses, *O*pportunities and *T*hreats.
2 The reasons for or causes of their occurrence.
3 The possible actions which could be taken.
4 The most appropriate course of action.

This analysis is often carried out in respect of (a) the organisation itself and (b) the environment in which the organisation operates.

Once a business plan has been produced arrangements are made for monitoring it, reporting on progress and revising it as appropriate. Quarterly reports are common practice. Responsibilities for monitoring performance must be assigned to officers and arrangements made that channel the information to key decision-makers on time and in readily understood formats.

For the increasing number of local government services which are subject to competition from private contractors, it is also necessary to adopt a market strategy which involves an analysis of customer needs and preferences, the competition, pricing strategy and market share.

Business planning is not a mechanical process. Although performance indicators demonstrate whether the plan's objectives are being met, complexity and uncertainty will always be features of the organisation's internal and external environments, making control in a mechanistic way impossible. Jackson and Palmer (1992: 3) suggest that the feedback loop from performance indicators to strategic decision-making can be viewed as a learning process in which 'strategy is adjusted, moulded and crafted as decision makers learn from experience'.

Old-style management used to concentrate on inputs rather than outputs in managing services. What mattered was largely managing budgets and staff. Although this remains essential, management today is increasingly concerned with outputs because these are what the users of services value. Outputs are tangible benefits to service users:

> We should remember that only something which has met a real customer need is an output. A street sweeping vehicle going up and down 100 miles of road is an input; a street sweeping vehicle removing all litter from 100 miles of road is an output. The processing of 200 forms is an input; deciding on what to do with each of 200 planning applications is an output. Making visits to 50 food premises is an input; making sure that the food on sale in 50 food premises is safe to eat is an output.
>
> (Tam, 1993: 54)

Although much work in public sector organisations is routine, they have had to cope with considerable change in recent years. This is particularly true of local government, where examples of major recent changes include the introduction of the poll tax and its replacement with the council tax, CCT, new responsibilities for community care, government requirements that local education authorities reduce surplus places in schools through closures and rationalisations, and local government reorganisation.

Project management is an essential method for managing change at a more detailed level than business planning. It involves programming the key tasks

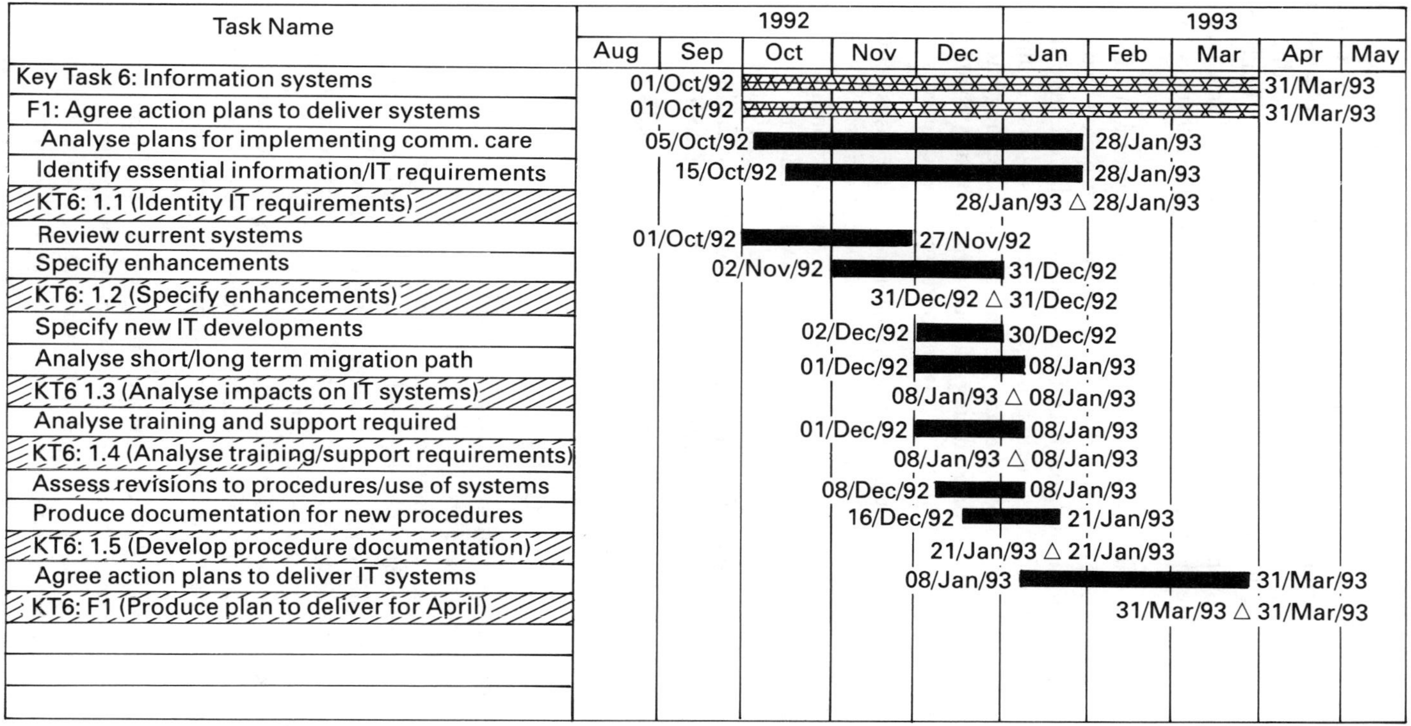

Task Name	Start	Finish
Key Task 6: Information systems	01/Oct/92	31/Mar/93
F1: Agree action plans to deliver systems	01/Oct/92	31/Mar/93
Analyse plans for implementing comm. care	05/Oct/92	28/Jan/93
Identify essential information/IT requirements	15/Oct/92	28/Jan/93
KT6: 1.1 (Identity IT requirements)	28/Jan/93 △	28/Jan/93
Review current systems	01/Oct/92	27/Nov/92
Specify enhancements	02/Nov/92	31/Dec/92
KT6: 1.2 (Specify enhancements)	31/Dec/92 △	31/Dec/92
Specify new IT developments	02/Dec/92	30/Dec/92
Analyse short/long term migration path	01/Dec/92	08/Jan/93
KT6 1.3 (Analyse impacts on IT systems)	08/Jan/93 △	08/Jan/93
Analyse training and support required	01/Dec/92	08/Jan/93
KT6: 1.4 (Analyse training/support requirements)	08/Jan/93 △	08/Jan/93
Assess revisions to procedures/use of systems	08/Dec/92	08/Jan/93
Produce documentation for new procedures	16/Dec/92	21/Jan/93
KT6: 1.5 (Develop procedure documentation)	21/Jan/93 △	21/Jan/93
Agree action plans to deliver IT systems	08/Jan/93	31/Mar/93
KT6: F1 (Produce plan to deliver for April)	31/Mar/93 △	31/Mar/93

Timeline: 1992 (Aug, Sep, Oct, Nov, Dec); 1993 (Jan, Feb, Mar, Apr, May)

Figure 5.4 An example of project planning the introduction of new information systems for community care

which make up a project, such as introducing new arrangements for providing care in the community implemented in April 1993. These changes meant that local authorities had to introduce new assessment procedures for individuals, plan care in residential and nursing homes, agree discharge arrangements with hospitals, consult with GPs and primary health care teams, introduce purchasing arrangements, implement financial and management support systems, and provide information about the changes to the public.

There are many other examples of projects in local government. They are self-contained pieces of work, with pre-determined end products or outputs. These outputs are realised through a set of activities that have a finite lifespan and a finite set of resources, such as people's time, money and equipment. Figure 5.4 shows an example of a project plan for implementing one of the key tasks necessary for introducing the community care changes. By planning a project in this way, the following outcomes are achieved:

1. Visible direction and purpose for all contributors to key tasks delivery.
2. An enhanced profile for a complex and urgent set of activities.
3. A mechanism to ensure controlled delivery of key tasks on time, within budget/resources, to specification and in a co-ordinated way.
4. Enhanced capacity to plan and schedule activities for individuals.
5. Greater accountability for completing specified tasks to deadline.
6. Opportunity for flexibility within an overall output-driven framework.

Business planning and project management are essential skills in today's public sector management.

CONCLUSIONS

A customer-oriented culture is a new experience for traditional parts of the public sector which have previously worked in an environment of rules and regulations. The reorientation towards customer management needs to be well-planned with staff fully involved. Public service workers have had to cope with many changes that have added to their workloads; it is therefore essential that managing services for quality is experienced as of positive benefit, especially for front-line staff. Skelcher (1992) identifies a key role for managers in this respect, encouraging and supporting employees' initiatives and improving the quality of relationships within the organisation. He argues that employees are themselves the customers of managers' skills and resources.

Business planning should have the same role as a strategic means of releasing the potential of staff to deliver quality services. It should be the prime expression of a public sector organisation's culture, with staff development linking employees to its values and objectives. Whilst the business plans of private firms have the underlying purpose of achieving profits for the company, public sector business plans will contain the overall values which derive from the purpose of providing public services. These values are based on moral and political positions. They often place restraints on action. For example, service improvements or spending reductions may be required to discriminate positively to give priority to where needs are greatest.

Although public sector business plans are about providing an efficient and effective service subject to such restraints, commercial considerations are becoming more important. This is especially the case for those direct services such as housing repairs and management, street cleaning or leisure management which are subject to compulsory competitive tendering and are required by law to make a specified rate of return on capital. Many local councils have responded by turning their services in these areas into effective business organisations.

Quality management is particularly important in public services because failure to deliver satisfaction to the user is not signalled by people being able to 'shop elsewhere'. There have to be other mechanisms in place to keep pushing services to maintain quality standards: what might be termed *customer feedback mechanisms*. These include decentralisation to improve access and responsiveness, user and non-user surveys, 'before and after' surveys, opinion polls, consumer panels, suggestion boxes and regular analysis of complaints.

It is sometimes suggested that because users of public services are not in a market situation they should not be called 'customers'. There is debate about this, but it is now a common term in the public management literature. Tam explains the thinking behind this as follows:

> In the context of local government, the reason why we can legitimately expect local people to follow our resolutions and policy directives is because in turn the local people can legitimately expect us to meet their needs in a number of areas. This makes them, in the most important sense of the word, our *customers*.
>
> (Tam, 1993: 2)

However, employees in public services have to relate to the public in different ways. Not everyone is a 'customer' or service user – many are parents, carers and people on waiting lists. Frequently, what someone wants cannot be provided because of rationing processes such as waiting lists for council houses. Just because needs increase – a waiting list gets longer for example – local councils are not in a financial or legal position to fund increases in services. Because most public services are paid for collectively through taxation, users are not individual paying customers. It is society at large which pays and which therefore has an overall interest which is represented through the democratic system. As already noted, many consumers of public services are compelled to use them. Furthermore, most consumers do not have choice between different suppliers of a particular service. Even with compulsory competitive tendering, the consumer usually has no choice but to use the provider who wins the tender. As Dawson writes:

> The many relationships which an authority has with its citizens makes it even more important to take consumer care seriously. The job of treating people fairly and courteously is even more of a challenge when consumers are not able to just buy what they can afford, or if they are not satisfied, go elsewhere.
>
> (Dawson, 1992: 13)

Managing for quality is part of a series of changes towards 'user friendliness' stimulated by criticisms of public sector bureaucracy and by higher public expectations. There is consensus about the need for improvements in standards of service and accountability to the public.

As earlier chapters have discussed, there is on the other hand no consensus about the extension of competition and privatisation into public services, and considerable argument about whether these measures are damaging or enhancing quality. The pursuit of quality is about high standards for public services that are based on what the public want. The underfunding and de-democratisation of public services in the UK is therefore contradicting other efforts to improve the quality of services. Quality has often become the responsibility of a 'new magistracy' of appointed decision-makers in the NHS, grant-maintained schools, further education colleges, housing agencies and other local levels of the public sector (Gyford, 1993). Although there is now greater emphasis on quality assurance, there has been a loss of local democratic control over the local public sector:

> the removal of functions from multi-purpose local government to single service agencies, no matter how localised, may clearly reduce the previously existing scope for 'horizontal' political debate about both overall and relative levels of service provision and funding whilst promoting the vertical accountability of service agencies upward to the relevant Secretary of State. It is a curious irony to see such principles of democratic centralism being applied in a post-Communist world . . .
>
> (Gyford, 1993: 77)

Managing for quality should not be confused with the 'management culture' which is being established in British public services by central government. The latter can be traced back to the Financial Management Initiative (FMI) in the civil service launched by the Conservative prime minister Margaret Thatcher in the early 1980s (see Chapter 3). This strengthened the accountability of public sector managers by making them responsible for performance against financial and service targets. The national citizen's charter is a logical extension of this new managerial ethic

(Doern, 1993). Standards and budgets are largely prescribed centrally rather than decided locally, effectively taking policy out of local politics.

Whilst national standards and entitlements are appropriate in many areas of service provision, the ability to respond to local needs is essential to successful urban policies. The methods of decentralisation, customer management, charters, policy evaluation, business planning, customer contracts, service level agreements and project management all assist with responding in this way. The balance between centralisation and local control has swung dramatically towards the centre in recent years, so these methods have become tools of central government policy implementation. However, this should not discredit them as potential tools of local democracy.

6

COMMUNITY DEVELOPMENT

Although urban policy is about achieving changes that benefit the welfare of local populations, it will not be able to fulfil this purpose if people cannot voice their needs or take action to win change themselves. This chapter describes how community development has assisted with this process by bringing the many different interests represented within multi-racial and multi-cultural communities into the co-ordination and planning of services. It reviews the different aspects of community development, its aims and methods. It considers its role in the local public sector, from individual projects to corporate strategies, and concludes by assessing its future as an approach to urban policy.

Community development is a strategy for putting urban policy into practice. The Association of Metropolitan Authorities (1993b) advocates a central role for it in local government, and other agencies such as health authorities have in recent years shown increasing interest in a community development approach.

In essence, community development is about people taking action together and developing the knowledge, skills and motivation to express their needs and improve conditions, either in a particular geographical neighbourhood or for a particular 'community of interest', such as black people or young people. This sphere of activity is sometimes called the 'community sector'. It ranges from people helping their immediate family and neighbours to major projects or campaigns.

Community development is a community-based approach to urban policy which supports joint working and partnerships with people in their neighbourhoods or communities of interest. By facilitating participation, community development supports policies for decentralised services and public participation in decision-making. Under compulsory competitive tendering, it has a new role in supporting community-based organisations to tender for contracts, bringing services and jobs to areas of high unemployment.

For local government, community development is, 'a way in which a local authority deliberately stimulates and encourages groups of people to express their needs, supports them in their collective action and helps with their projects and schemes as part of an authority's overall objectives' (Association of Metropolitan Authorities, 1993b: 10). It builds on the democratic traditions of local government, although it is an approach also used by other public bodies such as health authorities, training and enterprise councils and urban development corporations. It will be ineffective, however, if different public bodies pursue their own strategies. Community development cuts across agency boundaries; its focus is on problems such as crime or lack of jobs rather than the separate responsibilities of different organisations. Local authorities are in the best strategic position to involve the local public sector in community development strategies on a partnership basis which also needs to include voluntary organisations and the private sector.

Community development is about strengthening representation and participation where there are barriers to this; an equal opportunities framework is therefore essential and this requires the investment of resources. Giving voice to people who are disempowered by disability, caring responsibilities or lack of English means resourcing their participation: using staff who can communicate effectively with target groups, providing access and facilities for disabled people, and providing child care, travel expenses and translation.

DEFINITIONS AND AIMS

Community development implies that members of a 'community' have common needs. It could be argued that if there are not common needs there is not a 'community'.

Much has been written about defining a 'community' and there are now probably well over a hundred alternative definitions (Coombes, 1993). 'Communities' might broadly be defined as local populations identifiable from their wider society by a more intense sharing of concerns (a *community of interest*) or by increased levels of interaction (an *attachment community*). Table 6.1 summarises some recent approaches to the concept, distinguishing the different bases of community along three dimensions of geography, behaviour and identity.

There are two main types of community: territorial and non-territorial. Territorial communities arise from the sharing of a neighbourhood and place-bound interaction. They are often defined by a community geography of travel to schools, shops and other services. People appear to identify most often with quite local areas, although this may still be as large as a whole town, and most people's sense of territorial identity or 'home ground' is not well defined (Hedges and Kelly, 1992).

Non-territorial communities are based on shared identities which are not place-bound. Britain's 1.5 million Muslims are a community in this sense. Increasingly, issues of identity in respect of race, culture, religion, gender and

Table 6.1 Some recent approaches to the concept of 'community'

Davies and Herbert, 1993	*Hedges and Kelly, 1992*	*Coombes* et al. *1992*	*Dept of the Environment, 1992*
Areal content		Landscape Economy Infrastructure Facilities	Topography Industry Transport
Behaviour (interactions)	Familiar territory Social contacts	Voluntarism Institutions	Personal mobility
Identity (perceptions)	'Home ground' Roots/ancestry	Heritage Demography Culture	History Demography Sport, leisure and culture

Source: Coombes, 1993.

sexuality have become the basis for community action, community service and community work.

Communities are diverse and complex. Britain's 'black community', for example, is in fact many communities defined in terms of racial, social, cultural, religious and class differences:

> The traditional definition of a community being identified by geographical proximity, or even by a shared culture, serves little purpose when trying to address the distinct and specific community development needs of Black people. Black people as a group have a shared experience of white racism. But this simple fact must not be allowed to conceal the many differences that exist both between and within Black communities ... The implications of this are that different ways of regarding and servicing communities need to be developed.
>
> (Husbands, 1993: 1)

Identity and interest 'communities' are emerging with the growing fragmentation of society caused by a greater diversity of economic, social and cultural experiences than in the past. Although there is greater potential for social division in this situation, there is also potential for alliances and the sharing of experiences. For example, black community work has recently looked to the African National Congress in South Africa for a model of political education. The ANC, in its transition from a liberation movement to a political party, is using community education to raise popular awareness about what it means to vote, using drama, literacy projects and voter registration campaigns.

Although communities of interest have become of growing importance in British society, geographical neighbourhoods continue to be important in the lives of many people, particularly women and children, older people, unemployed people and people without cars. These are often priority groups for urban policy and public services generally, so community development is of particular relevance in focusing action at the local level where there is shared dependence on local services, often shared threats from problems such as crime or debt, and a shared desire to improve facilities and conditions.

'Community development' encompasses geographical communities and communities of interest. Widely defined, it is pursued through area and neighbourhood forums, the introduction of user-control of services, and the devolution of control over policies and budgets to forums such as the race and women's committees established by many local authorities in urban areas. In practice, it can take three main forms: community action, community service and community work.

Community action is usually about fighting an issue either to oppose a proposal (such as a local school closure), or to press for resources (such as a community centre). It may also involve people pressing for change in how services are delivered; greater participation in decisions made by local councils or health authorities is often a focus of action. Campaign tactics are used to tackle concrete issues, with the objective often a redistribution of resources and power to the 'community'.

Community service is used to describe voluntary action, such as local people running a mother and toddler group or a youth project, getting involved in a 'clean up' campaign or organising a community festival. The objective is to develop and run community-oriented services with involvement from the community. Statutory agencies are increasingly supporting community service as a way of targeting the delivery and uptake of services in 'deprived communities'. Examples include community health projects, family support centres, community policing and community businesses. In social services, community development has a role in strengthening the organisation and voice of users and carers, providing preventive support services, and in enabling community groups and smaller voluntary organisations to contract with the local authority to provide caring services. Such contracts are increasingly replacing grant aid as a means of funding voluntary organisations.

Community work is generally used to describe what paid community workers do. Community workers are employed by local councils, voluntary organisations, urban development corporations and other agencies to promote community development, either generally or with a particular focus. The Association of Community Workers sees the essence of the community worker's role to be, 'that of enabling people to achieve things themselves – through suggestion, education, organisation-building and providing information and advice' (Association of Metropolitan Authorities, 1993b: 11).

Community development strategies require the skills of specialist community workers. However, if they are to be corporate strategic approaches across all services then many staff who work with the public should have community work skills. The growth of community work, and with it a community work profession, has been partly a response to the failure of other professions to work with local communities.

Twelvetrees (1991) distinguishes between *community development* and *social planning*. The former is based on the principle of 'starting where people are at' and promoting *self-help*, either by providing a service through voluntary action or by campaigning for services and improvements. Community work in this context is 'non-directive'. It supports local people in what *they* decide to become involved with.

Social planning, on the other hand, involves liaison and joint working between local residents and local services. It is aimed at improving services through partnerships. Many local authorities have established *area strategies* in neighbourhoods where the council and local residents recognise a need to work together to tackle problems. These generally consist of committees which meet to bring together residents active in different community groups and professionals from different agencies working in the area. They agree needs, plans and priorities. There will often be sub-committees working on particular issues, such as young people, jobs and training, or health. The basic principle behind this type of work is to agree objectives together and to ensure that meeting needs identified by local people is explicitly reflected in the service plans and day-to-day work of service providers. However, the extent to which such forums are used just to consult local opinion, rather than genuinely to move towards community or neighbourhood control of resources, depends on how far the local council or other funding agency is prepared to devolve its own power. Such forums can take a long time to develop to a point where there is strong community involvement.

Community development as social planning is increasingly recognised as a way of involving people in local government (Tam, 1993). In 1989, the Council of Europe endorsed a Resolution on Community Development promoted by the Standing Conference of Local and Regional Authorities. It called for community development to be, 'high on the list of political options of local and regional government in Europe'. The concept of community development used is more wide-ranging than local action, or just local government services, and embraces integrated social and economic development, citizen participation and partnerships between different agencies and local communities. McConnell (1991) summarises the thinking behind the resolution as follows:

1 A recognition that communities should be consulted and involved in decision-making.
2 Comprehensive strategies by public authorities to ensure that community consultation and involvement permeate all parts of social and economic planning and service delivery.

Table 6.2 Aims, outcomes and methods of community development

Aims	*Outcomes*	*Methods*
To extend local democracy	Increased public participation in civic activities and self-help	Community education Financial support Technical assistance Decentralisation and accessible services Community work support
To empower disadvantaged groups to articulate interests To provide structures and resources to make participation happen	Increased ability of disadvantaged communities to articulate their interests and concerns	
To strengthen independent community networks and organisations	Strengthening of community organisations	
To promote community control over assets	Increased control of local communities over assets/greater sense of ownership	
To tackle problems and needs at local level more effectively	More effective targeting of limited resources to tackle needs	
To assist residents and consumers to have real power in their partnerships with the local authority and other agencies	Improved quality of partnerships between the local authority, other agencies and local communities More effective inter-departmental and inter-agency work at local level/multiplier effects Improved customer services	

Source: Developed from Association of Metropolitan Authorities, 1993b.

3 Providing access for citizens' groups to sympathetic professional assistance, technical advice and advocacy, facilities, money and resources so that they can participate effectively.
4 A change of attitude among professionals to assist them to be far more sensitive to the needs, concerns and ideas of local people.

Table 6.2 summarises the key aims, outcomes and methods of community development. The emphasis is on working against disadvantage, poverty and powerlessness. Although some projects seek to increase local incomes by creating jobs or maximising the take-up of benefits, the reality is that most work has to concentrate on supporting self-help and pressure group activities. Not having enough money to have a reasonable standard of living is made worse when homes are difficult and expensive to heat, shops are of poor quality with higher-than-average prices, public transport is poor and costly, jobs pay too little to meet the costs of travel or child care, and possessions are constantly under threat from the next burglary. These problems face many people living in urban areas with high rates of unemployment, but there is often a strong capacity in the community to do something about them. Community development aims to support this existing capacity, building training and employment projects on existing skills and interests, providing child care to enable women to realise their potential, tailoring housing and security improvements to local knowledge about what is needed, and tapping local entrepreneurialism with initiatives such as food co-operatives and credit unions which recycle money back into the community.

Community development, however, often involves conflict. Community workers can be in a contradictory position if it is the organisation that employs them which is the target of local campaigning. In these circumstances, community workers have to support the groups they are paid to work with, but have to be sensitive to what is likely to be acceptable for a public sector employee to do.

Organisations which employ community workers should have a community development policy which at least reduces the potential for conflict by stating clear objectives. These should be to make the organisation more responsive, open and accountable to the needs and demands of local people. But they are usually framed to support other policy commitments as well, such as targeting resources where needs are greatest and equal opportunities. They will often contain clear recognition of the role of community groups and voluntary organisations. Many make a commitment to providing services that are complementary to this role and in partnership with community-based groups and organisations.

STRATEGIES AND METHODS

The Association of Metropolitan Authorities (1993b) advocates that local authorities adopt *community development strategies* which are managed within a single department, preferably under the chief executive, and accountable to a high-level, inter-service committee. The development of these strategies involves a number of key stages:

1 Clarification of the views of elected members about community devel-

opment; production of a policy stating the aims of a community development strategy.

2 A review of existing relevant policy and strategy statements and the introduction of consistent community development principles into these policies and statements.
3 An audit of the range of departmental resources and other agencies able to support community development.
4 Compiling information about community groups, including a corporate register of contact people and activity categories.
5 A detailed needs assessment, consulting with local communities and community organisations. Identification of strengths, gaps and opportunities from this.
6 An audit of the community development skills and training needs of staff who have primary responsibility for working with the public.
7 Decisions about budgets, management arrangements and delivery mechanisms (including grant-aid to community groups and training), priorities and targets, and desired outcomes.
8 Monitoring and evaluation to check that the strategy is being implemented and to pick up on problems.

Community development strategies should encompass both ‘community services’, such as adult education or outdoor recreation, and mainline services, such as housing and social services. Three main elements are usually found in these strategies:

1 *Direct support to community groups*. This includes community work with groups to establish new services, to enable participation in the planning and management of services, to develop anti-poverty initiatives such as welfare rights campaigns and childcare projects, and to develop community and social facilities to support local participation, such as resource centres.
2 *Development of open-access, community-based services*. This includes establishing direct services to parents, young people and children, such as playgroups, childminding, playcentres, community centres, youth centres, adult education facilities, support groups, welfare rights services and care services.
3 *Promoting user-participation in service delivery*. Services which adopt a community development approach must be committed to service delivery which promotes access and involves users in decisions about services.

Involving the community in decision-making is a goal towards which many local authorities are working. North Tyneside Council, for example, has established neighbourhood consultative forums, involving local residents in all services delivered to their local area. In the 1980s, this council also established unemployment centres to provide social, educational and recreational opportunities for unemployed people. Their remit was subsequently broadened to include lone parents and low-paid workers, and they were renamed people's centres. Gallant comments about these centres that:

> Their key achievement has been to retain a campaigning dimension addressing structural issues while also responding creatively to the needs of people who have previously never been in a position to articulate their needs and aspirations or to have command over some resources. This has led to significant achievements in poetry and crafts, in sewing and sociology, that would never have happened in more conventional settings. But it has allowed ordinary people to progress from being learners to being teachers and from the immediate to the longer-term in their goals for self-development.
>
> (Gallant, 1992: 43)

North Tyneside Council is also using the Neighbourhood Initiatives Foundation's *Planning for Real* method to work with local residents on improvements to their neighbourhoods. Former mining village residents were involved in using a scaled model of their village to plan improvements, taking the model around various local venues. The work of council departments was then co-ordinated with the residents' ideas.

Planning for Real was used with community groups in the Meadowell estate, a scene of rioting in the summer of 1991. An action plan for a 'community village' was made, encompassing social and economic projects as well as housing schemes. A Community Development Trust has sought to bring together different community groups with council departments and agencies operating from outside the estate.

Although professional community workers are employed within separate service departments and by grant-aiding voluntary organisations and community groups, there are major advantages to be had from a community development unit with a corporate role in a local authority. This is then able to provide a visible lead role in establishing a community development approach across the organisation. St Helens Metropolitan Borough Council, for example, employs its community workers in a single team. Until recently, each worker had their own patch or neighbourhood together with a borough-wide specialism. However, the growing volume and complexity of legislation on housing, social services and education have led the council to move away from a neighbourhood or patch basis and to deploy its community workers according to specialisms in these areas (Association of Metropolitan Authorities, 1993b).

The St Helens team has short-term and long-term objectives which are set out in its action plan and are used to prioritise work. The core competencies that are needed, such as research skills and facilitative skills, have been identified and are used in job specifications, staff development and training.

Each community worker has a case load of groups or projects and is supervised through monthly meetings with their manager. These meetings review the month's work, analyse major issues that have occurred, and agree actions for the next month. Supervision notes are written up and used at staff appraisals, when opportunities for career development, needs for training and current performance are reviewed. Each community worker also completes a short quarterly report for each of their groups or projects, reviewing outcomes over the period against objectives. These reports are used to present a largely quantitative picture of the team's performance to its council committee.

Reviewing the performance of community development is particularly difficult because as much emphasis needs to be placed on how objectives are achieved (the process) as on the degree to which they are achieved (the outcome). In St Helens, this involves staff in demonstrating in their reports how the way they are doing their job relates to basic community development principles.

Community development is both an approach and a set of methods. 'How to do it' textbooks such as Twelvetrees (1991) are guides to methods which can be used to set up, work with, develop and strengthen community groups. The key skills needed for this work can be summarised as:

1 Research and information gathering.
2 Information sharing and liaison with different agencies.
3 Communication skills.
4 Group work skills.
5 Development of strategies and tactics to achieve aims.
6 Skill in running meetings.
7 Resource gathering, including money, premises and volunteers.
8 Training and education: passing on the community worker's skills to others within community groups.

The main focus of community development is organised groups in neighbourhoods and housing estates. Such groups include tenants' associations concerned with the quality of local housing provision and housing services; women's groups active in such issues as local health services or training and job opportunities; consumer co-operatives such as credit unions and local food shops; youth projects and childcare provision set up and run by local people; and projects providing support to victims of racial harassment or domestic violence.

An increasing amount of community work is not focused on small localities but involves work with 'communities of interest' or on single issues. This includes work with minority ethnic groups, women suffering domestic violence and homeless people. Examples include racial harassment support groups, women's refuges, and 'drop in' centres offering advice and support. Many urban local authorities have established race and women's committees, with devolved powers over policy and budgets. Examples of single-issue projects and campaigns include community arts such as street theatre, homelessness campaigns and health projects.

Youth work is another example of working with a community of interest. It is often an integral part of the community education service run by local authorities. Youth workers spend time with young people where they congregate, working as informal educators. A major problem they have to face is the alienation from school which can affect young people living in depressed areas where there are few prospects. They also run and develop projects, often designed to displace the activities of young people away from crime.

An example of single-issue work is community enterprise. During the 1980s there were many local initiatives to develop employment opportunities in neighbourhoods and estates with high unemployment and little chance of receiving major private sector investment. 'Community businesses' are a way of targeting such areas for developing economic activity, involving local people in owning and controlling a trading organisation. Community development support is targeted on people who have been long-term unemployed or are otherwise excluded from employment to help them to a stage where they can develop and run a business.

Community enterprise is often presented as one of the most valuable ways to release the potential of deprived communities. Typical activities include managed workspaces to house local businesses (sometimes on a subsidised basis), food shops, recycling and refurbishment projects, printing and publishing, and credit unions (see Chapter 8). Local authorities often provide loans, counselling, business planning and marketing advice to these initiatives. Such support is extremely important, as one of the main reasons for their failure is management problems and a lack of harder-edged commercial and business development skills (McArthur, 1993).

Community economic development trusts are an extension of the community business idea. These are umbrella organisations which aim to develop projects or jobs in priority areas. An example is the Manor Employment Project, a community economic development project set up in an area of mass unemployment on Sheffield's Lower Manor Estate (Pedlar *et al.*, 1990). Led by two professionals, the project converted a disused council works department depot into cheap workspace for co-operatives and community businesses. The precarious nature of these economic initiatives, however, is illustrated by this example. All the early businesses foundered, and only businesses that came later in search of cheap premises survived.

Community groups and projects will often be based in a local building, such as a community centre, which acts as a meeting place and provides resources such as office facilities, childcare facilities or a cafe. Community buildings are also often used as a base from which local government or health services staff provide local services, for adult education classes and for advice services such as welfare rights. But just as community work is not only the prerogative of community workers, the community use of buildings should also apply beyond 'community centres'. For example, local authorities are able to direct school governors to open schools for use outside normal school hours for community purposes. The full cost of this use, however, has to be paid for by the user or local authority.

There are a number of examples of 'community networks' being formed to bring together community groups around their common interests, purposes and problems. These can be permanent, such as the Bristol Community Groups Network (BCGN), or temporary, such as the network of Belfast community groups which formed to influence the Belfast Urban Area Plan (Blackman, 1991). Community networks not only give collective strength to their individual community groups but also constitute forums with which local authorities can consult. Bristol City Council, for example, consults with the city's forum of tenants' associations on all new policy papers before they go to the council's housing committee for decision.

POVERTY AND EMPOWERMENT

The local focus of much community development reflects the fact that deprivation concentrates in certain areas. The characteristics of these areas often themselves influence needs for public services, such as high crime rates, lack of shops and poor environment, producing a multiplicative negative effect on the quality of life of local residents.

Deprived areas are now overwhelmingly areas of council housing or housing rented from housing associations (see Table 6.3). For financial and tax reasons, owner occupation has become the dominant way of people housing themselves in the UK, leaving in the main only those who are not in a financial position to take on a mortgage to depend on renting accommodation from a local council or housing association. The size of the council housing sector has shrunk since the 1980 Housing Act introduced the tenant's 'right to buy', resulting in the sale of the better parts of local councils' housing stocks. The overall effect has been to concentrate poor households, especially younger poor people, in the worse parts of the council housing stock. Many estates have effectively become welfare ghettos.

Table 6.3 Average income of social housing tenants as a percentage of the national average, 1981–90

	1981 (%)	*1990 (%)*
Council tenants	73 (combined)	48
Housing association tenants		45

Source: Page, 1993.

Many council estates are experiencing major social problems caused by unemployment, debt, high child densities, and poor security and safety. Women in particular face a struggle to make ends meet and bring up children under these conditions. Discussions groups held with mothers on a 'problem estate' in Newcastle revealed how those who had moved to the estate from other areas noticed a real difference in their children following the move:

> It's totally different down here – I moved from Durham and down here I just cannot control them. They're like – changed. I mean, there's nothing for them to do, and they're chuckin' eggs off pavements, knockin' on doors. There was plenty of baths and all for them there, there's nothing here . . .
>
> My first two weren't born here, and where we were before they got the chance to meet nicer kids. But my last ones were born here and them two are right little villains. They picked it all up. But I was still the same parent to them. I brought them up the same way. And I knew it wasn't anything different I had done, it was just society . . .
>
> (Hill *et al.*, 1991: 23)

Accounts like this are fairly typical of the problems that have emerged on many housing estates over the last 10–15 years. It is often the case that people want to leave, creating a spiral of decline and voids in the housing stock, and leaving behind only those with no other choice. Sometimes there is no other option but to demolish housing that declines in this way. But in other areas there are in fact strong attachments to the locality despite its problems, which are often perceived as resulting from a minority of 'problem families', little for kids to do, and negative council and police attitudes towards the area.

Local government is dominated by professional groupings such as housing managers, planners, social workers and teachers. These professionals have significant power as 'experts' in operational control of their services. For poor people, who most depend on these services, this power can be experienced as control exercised by others over their lives. Middle-class experts and their clerical staff act as 'gate-keepers' of resources such as housing repairs, social security benefits, health services, examination achievement and social care. Pahl (1970) termed such professionals 'urban managers', but this concept was widely criticised for exaggerating their power (Taylor, 1985). Whilst to some extent controlling 'who gets what', the immediate managers of services have little say in how much there is to distribute in the first place. Nevertheless, a division frequently exists between the residents of deprived areas dependent on the local state for services and the professional officers who run these services.

Local people living on the Newcastle estate described above expressed this situation as follows:

> People that live outside the area, they don't know what it's like to live

Table 6.4 Social residualisation of a Newcastle council estate: socio-economic group of the economically active population

	*1981 (%)**	*1986 (%)*†	*1991 (%)**
Professional/managerial and intermediate professional	27	3	2
Junior non-manual	33	30	15
Skilled manual	20	32	26
Semi/unskilled manual	20	35	57

Sources: *Population census; †Newcastle City Council 1986 Household Survey.

> here, sometimes they come here and patronise you ...
>
> When the professionals go into their jargon this puts the local people off, so therefore they lose them on the wayside. And then you'll find that everything local people want also goes by the wayside cause there's no one there to promote it and fight for exactly what they want on their estate. Everything seems to be decided by the professionals and people on the ground don't get the chance to say what they want ...
>
> (Hill *et al.*, 1991: 18–19)

This division between local residents and 'professionals' who do not live in the area reflects the social residualisation of council housing estates which were once more socially mixed. The Newcastle housing estate is a striking example (see Table 6.4).

COMMUNITY DEVELOPMENT IN ACTION

Much community development consists of projects which have evolved over time rather than being part of any strategy. Examples include after-school clubs, summer playschemes, tenants' associations, drug awareness groups, crime prevention projects and adult education groups. Local people may have pressed for statutory agencies to provide their services locally, such as basing an employment officer in a community centre to notify job opportunities directly to local people, provide routes into training and further education, provide help following training, and provide a local point of contact for benefit problems. This may then develop into work with local groups to build up skills and confidence.

In some cases, different community projects have come together to form community trusts. This enables projects to share support workers and have a stronger voice. In Newcastle, the Cruddas Park Community Trust was also a means of building a common purpose among local residents: 'Local people ... recognised the division within their own community and saw the group as a way of bringing the estate closer together' (Cruddas Park Community Trust, 1991).

Many community groups start very small and build up gradually through both personal contact and by holding meetings. Community work has often been essential, with much time spent listening to local residents and door knocking to get local views. Meetings may begin very informally in people's houses. After time, a committee may be elected to take forward the areas of concern local residents identify.

Some community groups form for a particular purpose, such as to oppose a school closure, and then decline and die once the issue has been won or lost. Others seek to become established and to have a permanent negotiating position with, for example, the local council's housing department. A community worker may need to put in intensive work in the early stages of such groups when they are finding a footing both in the community and with statutory bodies. Later on, the community worker will often withdraw from intense involvement and adopt a routine servicing role, or withdraw altogether.

Community education has an important role in building confidence for people who were not able to achieve at school, providing opportunities for part-time learning and often acting as a springboard for local people to move on into further or higher education, or other community projects such as a community business. In 1991, there were nearly 1.3 million enrolments at adult education centres in England (National Commission on Education, 1993). These are mainly funded by local authorities, although support has declined due to the increasingly stringent control of local government spending and the necessity to charge full costs, although often subsidised for people receiving benefits. The National Commission on Education (1993) urged greater support as part of an overall strategy to raise the level of education and training in the UK, whilst acknowledging the special role of continuing education in personal development, promoting mutual understanding, supporting health promotion at community level, and in social and economic regeneration projects.

Community education seeks to extend learning opportunities to adults, particularly to those groups who are most disadvantaged, so that life-long learning opportunities are available. It works on the principle of meeting people's expressed needs for learning, drawing on and extending the experience of participants.

Community education takes place either in formal education establishments or in venues such as community centres. It often links together informal education in community settings with opportunities to access formal education in colleges and universities. This is being encouraged by the development of more flexible and accessible learning opportunities by colleges and universities, such as access courses, modular and part-time courses, and accreditation of prior learning or experience.

Community education is an example of the general community development approach of rooting services within the community. This has been the origin of a range of community projects in British towns and cities, from community launderettes to after-school care. Such projects are of value in themselves, but they can also have wider impacts by, for example, helping to reduce high turnovers of tenancies on housing estates or reducing the length of unemployment experienced by local people (Cruddas Park Community Trust, 1991).

These wider benefits have been particularly evident in the case of tenant participation in housing. In recent years, local councils have provided a variety of opportunities for tenants to participate in the management of their estates. This has focused on 'problem estates' where the level of unpaid rents, empty houses and vandalism has prompted a community-based response. These estates are areas of poverty and low employment, high dependence on benefits, and higher than average health problems. High numbers of social workers per head of population are common. They are usually stigmatised areas, although this image contrasts with survey findings which show that often the vast majority of residents are positive about their estates despite problems that they want tackled (Holmes, 1991).

The Department of the Environment's Priority Estates Projects in England and Wales promoted the decentralisation of housing management and the introduction of tenant participation in the 1980s. Local authority housing departments established local estate offices with responsibility for lettings and repairs. Residents were actively involved in the development of the projects and the running of the estates. Evaluations of the projects revealed increased tenant satisfaction, reductions in voids, rent arrears and turnover, and overall improvements in economy and efficiency (Emms, 1990).

The Priority Estates approach was adopted more widely after 1985, when an initiative called Estate Action was launched. Security measures such as concierge and entryphone systems and physical upgrading have been prominent elements of Estate Action packages, based on local priorities. On some estates with high unemployment, Estate Action has promoted Community Refurbishment Schemes which finance environmental improvements and repairs using local unemployed residents. Estate Action, however, has been strongly linked to central government policies of expanding the private sector by transferring refurbished housing stock into owner occupation and 'diversifying' the area. Housing associations also often acquire council property for rent. Consultation with tenants is required before Estate Action bids by local authorities receive approval from the Department of the Environment.

Many local councils have adopted formal policies on tenant participation. Glasgow City Council adopted such a policy in 1984 and has one of the most progressive approaches to tenant participation of any local authority in the

UK, linking housing objectives to a wider community development strategy. There are six main elements to its programme:

1 A community renewal service which involves tenants as active partners in the upgrading of their homes and local environments, from planning to estate management agreements.
2 A tenant management co-operative programme, delegating the main housing management functions to tenant committees.
3 A community ownership co-operative programme, whereby both ownership and management of ex-council housing is passed to a membership-based co-operative.
4 An estate action programme, building participation into the delivery of services such as cleansing and recreation as well as housing.
5 An independently financed package of support for the city's tenants' organisation. This includes a tenants' action fund and a city centre tenants' resource unit staffed by workers directly employed by the Glasgow Council of Tenants Federation.
6 An area-based management structure in the City Council, enabling direct and local participation within the political machinery of local government.

As well as adopting a formal policy on tenant participation, it is important that participation is seen to be done at street level. In the best examples, this includes facilitating participation at every stage of housing renewal schemes, supporting tenant management co-operatives and community ownership, multi-agency working beyond immediate housing concerns to crime, ill-health and unemployment, and independent financing of an organised tenants' organisation which can employ workers directly.

Most tenant participation supported by local councils is still quite modest, with questionnaire surveys and public meetings being most common. Many housing authorities have adopted tenants' charters which inform tenants of commitments to respond to repair requests within certain time-scales, of the level of service they can expect, of rights to information and consultation, and of how to complain (see Chapter 5). These individual 'rights' are not the same as supporting tenants collectively, but they are important in treating tenants as customers of a service.

Like all community action, tenant participation may not have a strong footing in the community. Many tenants are overburdened by financial worries and other problems. Much often depends on having one person, or a small group of people, working energetically in their neighbourhood. Groups can fold when a key player gets a job or moves away. There is also a danger that gains by a tenants' group in one area are seen to be at the expense of tenants in other areas. People identify with very small areas and this can mean that community groups are seen to be dominated by those from one part of an estate where most of its members come from. This means that the role of the local council in allocating resources according to need remains essential. Community development is a means of tailoring these resources to local priorities.

Sometimes fairly structured techniques are used to decide priorities. Ritchie (1992) describes a one-day tenants' workshop which was organised to work out priorities for a modernisation scheme on the estate and a capital finance bid to the Department of the Environment. Role playing was used to explore the advantages and disadvantages of different approaches from the point of view of tenants, the council and the Department of the Environment (which approves funding for schemes). In this way, the different criteria that could be used to judge schemes were identified. A final exercise involved everyone in putting different coloured stickers on flip charts to find out what priority they attached to the different criteria. The method was found to be effective and fun, and was used in four subsequent public meetings:

> Crucially, the exercise showed that the main criteria for any bid were that spending should be directed to houses in greatest need of repair and should also be spread as wide as possible. This contrasted to current Council policy of carrying out WHI (whole house improvements) on geographically defined areas on the estate.
>
> (Ritchie, 1992: 18)

COMMUNITY DEVELOPMENT PRINCIPLES

Urban society in the UK is differentiated and multi-cultural. As a result, there is no one way to 'do' community development. However, Donnison *et al.* (1991) conclude that there are several principles which are to be found in successful examples. These are:

1. The needs of the whole community, including its most disadvantaged and marginalised members, must be taken into account.
2. Community development should involve finding the talent, information and knowledge that already exists in the community.
3. All stakeholders should be involved, including local councillors and private businesses.
4. Strategies need to be fully owned by the key leaders and actors in the local community.
5. The starting point for a community development strategy is self-assessment. This involves identifying needs, objectives and desired outcomes. A SWOT analysis is a useful tool. This involves asking what are the Strengths, Weaknesses, Opportunities and Threats for the local community?
6. Paid community workers may well be needed, but they should not specify objectives for local people, only help with the means for

achieving the objectives people decide for themselves.

7 There should be consultation about tactical choices as to which issues to tackle first, by whom and to what end.
8 The overall emphasis should be on learning and transferring skills. The aim of this is to ensure that the management of new initiatives and projects does not require constant intervention by professionals.

Although the shape of community development will depend on local circumstances, Donnison *et al.* (1991) also suggest some initial steps to follow in starting a community strategy:

1 Establish a steering group with a clear remit and with relationships sorted out with existing organisations.
2 Follow this with an intensive and highly consultative local research exercise to gather information about needs and ideas for improvements.
3 Develop short-term actions at an early stage to get the ball rolling.
4 Use public meetings to make presentations about the way forward.

Much community action depends on one person, or a small group of people, working energetically in their community. These people have often moved through a series of experiences, perhaps beginning by participating in a mother and toddler group, moving into housing action and then perhaps on to get involved in employment projects. A rare combination of skills is required, bringing together community development, entrepreneurialism, an understanding of how statutory agencies work in order to influence their decisions, and a strong commitment to the neighbourhood and its residents.

Problems can arise in community action when a key player finds employment and has less time to devote to community work, or when developments are pushed too rapidly by professionals, exceeding the capacity and skills of local people at the time. Filling paid posts in local community projects can be very difficult as a result of small-group loyalties and vested interests.

Much can be learned from good business practice. For example, community groups often do not work with objectives, targets and measures to track progress towards their objectives. However, such practices need to be introduced carefully. The danger of 'professionalising' community development and alienating people by foisting ideas on them must be avoided. This means continually reinforcing areas of confidence, always relating to familiar matters, and making everything explicit. Holmes (1991: 9) stresses that skills and competencies need to be acquired gradually, avoiding the situation where, 'quite often people are asked for a view and straight away expected to chair and do the minutes'. She identifies six main factors which put local people off involvement in community development:

1 Feeling that they would not be effective.
2 Experiencing failure.

3 Low self-esteem and lack of confidence.
4 Not wanting to fuss.
5 Not liking the other people involved.
6 Domination of 'experts' and 'professionals'.

It follows from this that people will be motivated if they feel they will be effective, have experienced success, see peers involved, feel some personal benefit or face a serious threat to their welfare.

Public sector organisations with a policy commitment to community development also need to be guided by 'good practice' principles. Many organisations now use community development to support a partnership approach with local people, and it is particularly important that people understand what is going on when they are invited into partnership with statutory agencies. The five 'Is' of involvement are a useful guide (Association of Metropolitan Authorities, 1993b):

1 *Information.* Are people allowed to know everything or will there be minutes, reports or other documents which remain confidential?
2 *Independence.* Can people get access to independent expert advice to help them interpret specialist information and to express their own opinions on specialist issues such as building plans?
3 *Initiative.* Can community groups develop their own ideas and plans, irrespective of anything the council might come up with, or is their role only reactive?
4 *Influence.* Will plans, time-scales and budgets be debated within the council only, or will people be allowed to participate in decision-making?
5 *Implementation.* Once decisions have been made, how far will people be allowed to participate in implementation and supervision?

In recent years, a number of new approaches to community development have emerged. These include:

1 Short courses and training programmes. Examples include the new Certificate in Tenant Participation, 'organising your community' courses, training for tenant members of Estate Management Boards, and training about contracting.
2 Arts techniques, usually aimed at building confidence and aiding expression. The techniques include socio-drama, oral history projects, 'instant books', storytelling and visual arts.
3 Organisation and management development techniques. These include SWOT analysis, operational research (OR), and cost-benefit analysis (CBA).
4 Technical aid. This is the provision of independent advice and assistance which enables people to deal as equals with experts. Examples include Planning for Real and other design projects.

5 Information technology. Community groups are acquiring personal computers to run databases and produce newsletters. Techniques such as Priority Search have increasingly been used with community groups, using a computer program to help identify the priorities of a group.
6 Contract community work. This new development involves a community worker contracting with a community organisation to assist with achieving specific objectives within a given time period.

These new ways of working are innovative and there is a demand for them among community groups. However, the danger with them is that they are often short-term, facilitated by consultants or specialists, and may be regarded by managers and policy-makers as community development, without the necessary commitment to long-term community work.

CRITICISMS AND CONCLUSIONS

One of the most frequent criticisms of community development is that it is based on the premise that problems lie within the community or the individuals who make up the community. The problems are those of 'deprived areas', and people living in these areas have 'special needs'. This has been described as the *social pathology* model of deprivation (Armstrong, 1982). In this context, community development is a process which:

> aims to bring about changes in the functioning of individuals, groups and 'communities' by facilitating their integration into more coherent wholes. The 'community' is assumed to be homogeneous in its needs.
> (Community Development Project, 1974: 4)

In 1969, the British government established twelve Community Development Projects (CDPs) as part of the first national initiative to tackle inner city problems. Researchers attached to the projects produced a number of reports during the 1970s which emphasised that these areas and their residents were in fact victims of external economic and political decisions. They used a *structural conflict* model to understand the situation of inner city residents in terms of social and economic inequality. Structures of inequality existed because society was based upon the private ownership and control of industry and capital. The analysis of the problem which the CDPs started with, and the possible solutions, were reappraised:

> To rectify the situation requires fundamental change in the distribution of wealth and power. The contribution which a programme like CDP can make is to work with local people in generating a political awareness of these processes, and support action which works towards change; in short a political education programme.
> (Community Development Project, 1975: 2, quoted in Armstrong, 1982: 32)

The strategy which developed was one of generating among inner city residents a political awareness of inequality and linking community action with the wider labour and trade union movement. The British government acted to suppress this by closing down the projects and restricting circulation of their publications (Lawless, 1979).

The significance of the CDP analysis is that the language of 'community' and 'meeting needs' was replaced with 'working-class' and 'interests' (Armstrong, 1982). The analysis has since been criticised for neglecting the fact that community action is not only about class interests. It is particularly about women's interests, because the sexual division of labour makes the home and the neighbourhood more of a woman's sphere. At the same time, the dominance of male interests in society means that many women are not only confined to this sphere by being denied the same employment opportunities as men, but also that women's concerns about the quality of the home and neighbourhood are seen as of secondary importance in economic and social policy.

The practice of community development is still very much place-bound, responding to the reality of social geography. This geography is created by uneven economic development, the perpetual shifting of production from one product line to another and from one place to another in search of better returns on investment. The accumulation of capital is a process which occurs at a world scale because large firms and financial markets operate globally. The local effects are the development and expansion of some areas and the decline and abandonment of others.

In his book *Beyond the Inner City*, Byrne (1989) argues that industrial decline has 'disorganised' the economic base of old industrial areas. A consequence of this has been the ghettoisation of working-class areas. Without the social order of industrial communities, civil society in these areas is also 'disorganised', with violence, crime and occasional riot. This leads him to argue that the state seeks to 'organise' poor people to counter the threat to social order they pose, and which is most apparent in the level and type of policing. However, Byrne also identifies certain types of community work, and of decentralised housing management using community work techniques, as methods of managing and controlling the poor. Priority area strategies, for example, can incorporate community groups into local council committees and introduce procedures they must follow to obtain funding. By enmeshing community groups in official structures and competition for grants, the potential for groups to take radical action, engage in critical political debate and question the unequal distribution of resources is suppressed.

This analysis is convincing although it tends to neglect the many examples of community work which have stimulated community and political action (Lees and Mayo, 1984). In the late 1970s to early 1980s, for example, a network of resource centres in several British cities supported local political campaigns about social security, employment, housing and planning issues. Byrne (1989) himself identifies the role of North Tyneside Council's people's centres in campaigns about benefits, job training and related issues. His argument, however, is that in North Tyneside the local council provided resources for community organisations to undertake campaigns that could be directed at the council itself (his main example being a successful campaign to press the council to boycott the recently introduced Employment Training Scheme which was regarded as an exploitative system of having to work for benefits).

Byrne's argument also underplays the effects of the last fifteen years or so of economic recession on community work. Community workers have found themselves increasingly working with people trapped in unemployment and low pay, lacking the skills and confidence to organise as active and militant groups. Political organisation has suffered from the defeats of working-class politics in the workplace and in local government. Community work has often retreated into approaches which focus on personal development and small-scale projects.

Community development, however, is experiencing something of a revival in the 1990s as part of the shift away from the bureaucratic organisation of services led by professional experts and towards user involvement. Woods (1991: 29) emphasises the need for community development to get, 'away from the concept of providing services for people to drawing up a programme of service delivery *suggested by them*'. An important consequence of this is that no one agency will have the means to respond. By promoting community development, local authorities can pursue an important enabling and networking role in their communities.

Like quality and decentralisation, community development can serve different interests. It can be about strengthening people's control over decisions and making these decisions more relevant to their needs. There is indeed a trend towards 'community practice' in public services which has this promise. But it can also incorporate poor people into welfare systems and self-help projects. Community development can be about social control with no empowering to challenge poor public services and sharp social inequalities.

Even the relatively modest radicalism of much community development is under threat from the new managerialism in the public sector. This type of work has promoted both self-help and the community defining its needs to press for change at local level. It has often had an emphasis on creativity and co-operation, with professionals working with community groups in non-directive ways. This model is constantly at risk of pressure to restrict public expenditure. Community development is usually a discretionary function for statutory agencies and is vulnerable to cuts. It is long-term, developmental and educational, and does not sit easily with the contract culture of performance indicators and value for money.

Community development is under increased pressure to deliver results. It is a vehicle for consulting with 'the community' to short timetables, for delivering programmes and projects, for bringing 'the community' into partnerships for urban renewal or health, and for delivering to contract in a situation where grant-aid is now often on a contract basis. Community workers are more likely to be employed at arm's length from the state where it is more difficult to influence policy and resource allocation. Church-based projects are increasingly prominent. Given these trends, community development may become increasingly dominated by welfare functions, with less emphasis on education and campaigning.

The proper role for community development in urban policy is to encourage comprehensive planning and joint working between agencies which is led from the bottom up. It is a holistic approach built around individuals, families and communities. This approach is particularly evident in black community projects. These have often had to start from scratch because of the unsuitability of traditional public services. Community development is about saying 'take this community as a whole' and then deploying resources accordingly. It is also about saying 'what are the strengths of this community' and then building on them.

By contrast, the norm in urban policy is to deliver services from different departments to 'clients' or 'customers' who are rarely considered in the context of their community and its strengths. In fact, it is rare to build on strengths: urban policy focuses on needs or weaknesses. Funding is often sought by demonstrating the extent of needs, and this in itself can be highly stigmatising. Community development approaches emphasise services which provide people with what they need to release their own capacities: for example, Finkelstein (1993) argues that disabled people are more interested in the provision of ramps, information in Braille and on tape, signing on television, and other means of accessing able-bodied facilities than in dependence on specialised welfare services administered by able-bodied people. Removing barriers to personal and social development, rather than creating dependency, is what the community development approach seeks to achieve.

7

URBAN POLICY AND RESEARCH

Urban policy has become increasingly complex, faced with many diverse needs, pressure on resources and expectations that outcomes are measured and evaluated. Research is therefore an important component of good urban policy. This chapter reviews its role and how it is applied in practice. It considers both quantitative research and the emerging contribution of qualitative research. Finally, the chapter discusses some of the issues which surround research in urban policy.

The main role of research in urban policy is to investigate urban conditions and to evaluate the effects of policies and services. Although much technical research in fields such as energy or pollution control is relevant to urban policy, this chapter is concerned with the increasingly important contribution of social research. Social research is used in urban policy to provide information on which to base decisions and to indicate the likely consequences of particular options.

Social researchers in public sector organisations are fortunate in working in increasingly rich information environments. Modern public administration involves very large flows and exchanges of information, mostly administrative data. Research, however, produces a particular type of information which is *meaningful* to policy-makers and service planners, and hence is often termed 'intelligence'. This can be produced by assembling, analysing and interpreting information extracted from administrative data to produce performance indicators or assessments of need, or it can involve primary research such as the collection and analysis of survey data.

A particularly useful contribution made by research is to anticipate change. Forecasting demographic change is particularly important, informing decisions such as the adequacy of the number of school places, the provision of health services or the zoning of land for house building. Increasing use is made of geographical information systems (GIS) to monitor and analyse patterns and trends in spatial data, such as Cross and Openshaw's (1991) crime pattern analysis system used by Northumbria Police. These systems are at an early stage of development but have the potential to provide 'early warning' indications of change, such as neighbourhood deterioration or incidences of health problems, enabling preventive action to be targeted.

Most research in government departments and agencies, local government and health authorities goes on behind the scenes. These organisations employ research staff but also commission research from market research companies, academics and independent research organisations. Bulmer *et al.* (1993) estimate that the number of graduates employed on quantitative social research in the UK is about 9,000. The two main employment sectors are market research and higher education, both with 29 per cent of the labour market. Local government is the next largest sector, employing 17 per cent of graduates working in this area. Within local government, research staff are mainly employed by housing, planning, social services and chief executive's departments.

Most market research agencies, and some independent research organisations, only supply research services to external clients. Most research units in local government provide research services only to the organisation of which they are a part. Academic research includes work for both external clients and an internal audience of other academics. The present chapter is not concerned with this latter type of research which seeks to contribute to scholarship and learning, although the government has recently placed greater emphasis in its funding of research councils for academic research on work that is relevant to the needs of industry and the public sector (Cm 2250, 1993). The chapter's main concern is with non-academic applied research for urban policy. This type of research is often undertaken within very short time-scales compared with academic projects. This means that research officers frequently have to rely on existing sources of data such as the population census, although high-quality primary research is also carried out (Canter, 1993).

Research is often part of a range of other activities carried out by professional officers. They will evaluate projects, review policies, monitor and plan services, investigate needs, develop information systems and collate, analyse and present information. The role of research as a specific professional activity is frequently underestimated. Researchers are rarely employed by education departments, for example, and in departments such as social services, where many research sections were set up in the 1970s, the number of research staff declined sharply as a result of cutbacks during the 1980s. Ironically, as Gostick (1993: 27) comments, 'this same economic climate emphasised the increasing need for better information systems, project and policy evaluation skills and value-for-money studies'.

In the 1990s, the introduction of systems for community care planning has seen social services research rise to a more prominent role again. The demand for research skills is growing in this and other areas where services are increasingly expected to be based on assessments of need and evaluated for their effectiveness and relevance to users. Examples include the use of skills audits and research into the factors that determine the success or failure of training and employment initiatives; research-based 'social audits' and 'community profiles'; the use of consumer surveys and focus groups to evaluate the quality of services; the development of environmental audits; and in health the new emphasis on research into lifestyle and socio-economic factors, and the cost-effectiveness of treatments. Local government finance is also a subject of research and evaluation by central government, local authority associations and individual local authorities, including studies of the distributional impact of local taxation and the use of regression analysis to develop standard spending assessments for the local authority grant system (see Chapter 4).

Overall, research and the analysis of information are becoming more important to the operations of the local public sector. However, there is a mismatch between the research skills which are available and those that are needed. Gostick explains the problem as follows:

> Research skills of this order are increasingly hard to find except for staff who have undertaken specific research training, and although a number of postgraduate research training courses are now available, these courses are still largely geared to the more academic research projects, with a predominant emphasis on increasingly sophisticated methodological and data processing skills. Whilst these are clearly desirable, researchers within local government also need to be equipped with more robust analytic skills; a project management orientation; an ability to work quickly and flexibly in the face of changing demands; and (crucially important) to have a well developed ability to present information in a clear and coherent way. Much local authority research never finds its way into the covers of a research report, but more often appears as a series of tables or recommendations in policy and practice documents. The ability to summarise and present complex issues clearly and succinctly is a key attribute for effective research in local government, and is a skill which is largely ignored by the majority of current research training.
>
> (Gostick, 1993: 28)

Gostick's comments apply both to the need for professional officers with the skills to undertake research and communicate meaningfully with both policy-makers and practitioners, and the need for other professionals to have the skills to commission and use research.

RESEARCH IN LOCAL GOVERNMENT

Research in local government covers a wide range of activities. It includes studies of social, economic, demographic, environmental and transport issues, as well as statistical research, operational research and policy analysis.

This work helps local authorities to plan the direction of their services and to base their plans on sound information and analysis.

The first local government research and intelligence unit was established by the 1963 London Government Act, creating a unit within the newly formed Greater London Council. The 1972 Local Government Act (which led to the reorganisation of local government in 1974) enabled local authorities to set up research units, although this was not a requirement. The new local authorities paid greater attention than in the past to research, but there was more emphasis on creating corporate planning units – as encouraged by the Bains Report (1972) on the management and structure of the new authorities – than on establishing research units.

Today, it is the larger councils which undertake the greatest amount of research and also tend to employ research officers directly, although all types of authority commission research from external consultants. The location of research within a local authority varies and it may not be clearly separated from activities such as policy planning or performance monitoring. Where an authority has a central research unit this is usually located in the chief executive's or finance department. Central units can also stand alone or be accommodated in a particular service department, often the planning department. Where there is no unit dedicated to central research, research is often a corporate function found in a central policy unit, performance review unit, or marketing and communications unit. However, researchers are often not working in a central unit but are instead in individual service departments, such as a housing or social services department. In the metropolitan areas, units funded jointly by several district councils are a common model. Such jointly funded units are less likely to undertake policy analysis (having no such remit in relation to any one individual local council) and tend to concentrate more on the provision and analysis of statistical information such as census and unemployment data.

The main forum for the exchange of information and promotion of research in local government is the Local Authorities Research and Intelligence Association (LARIA), which publishes a regular bulletin, *LARIA News*. Other professional associations include the Housing Research Group (HRG) and the Social Services Research Group (SSRG), which run seminars and produce publications.The Local Government Management Board publishes *Research Link* which seeks to improve communication between local government and academic researchers. The Board is a strong advocate of research, especially for management:

> Research can increase the capacity of local authorities to manage. Over the past decade there have been enormous changes in local government. A great deal has been uninformed by research, resulting in avoidable confusion and uncertainty. Yet the need to test and analyse new ideas cannot be overstated. Research can also play a critical role in anticipating

> emerging debates and preparing management approaches to meet the issues of tomorrow.
>
> (Clarke and Hasdell, 1992: 23)

Until fairly recently, local authority research was largely confined to population and traffic forecasting and to the analysis of population census data. This has fed into such areas as planning the provision of housing land, roads and school places. Despite the relevance of social research to the policy analysis and corporate planning functions of local authorities, there was rarely a major commitment to social research at chief officer level. The growing interest in research during the 1990s is a consequence of the following factors found throughout the public sector and not just in local government:

1. There is greater emphasis on assessing the need for services. This is being driven by pressure to target resources where they are needed most, and requirements to specify contracts with service providers. It includes establishing the scope of current services and comparing this with needs, thus helping policy-makers and service planners to decide what issues and changes to focus on.
2. The importance of having knowledge about consumer views and attitudes, both because less of the public are now willing to accept that 'the council knows best' and because consumer views are central to assessing the quality and performance of services.
3. The importance of performance indicators and the need to make sure that indicators measure meaningful aspects of a service and are placed in proper context, particularly comparing like with like. This helps to establish whether policies and services are achieving the effects expected of them.
4. Expectations that public sector organisations have a good knowledge base, researching where there is scope for change which will improve welfare, evaluating what really works and comparing the cost-effectiveness of alternative measures.
5. The increasing need for local authorities to know about conditions generally in the local area, including the adoption of enabling and influencing roles in relation to other organisations and government. There is also increasing awareness across all public sector agencies that needs should be assessed in context and are often part of a complex web of inter-related factors.
6. The increasing extent to which public sector organisations have to bid for money by making a case based on evidence.

Behind all of these changes is the shift in the public services from a bureaucratic model of service provision to a model based on contracting. Control by contract has raised two issues in particular which involve a very clear role for research. First, clarity about the needs and the standards which the contractor is expected to meet. There is growing interest in finding suitable measures of outcome by which to define services to be commissioned. Second, clear accountability in relation to standards of performance and outcomes. This requires monitoring, inspection and evaluation, with a role for research in either leading these functions or providing support.

The recent emphasis on quality has increased the importance of research into the public's views about services (see Chapter 5). Research is needed to assess the performance of services against their objectives and standards, to analyse the outputs and outcomes of policies and programmes, to investigate the perceptions of service users, to investigate unmet needs, and to establish employees' ideas and views. This means that public sector organisations need access to researchers with skills in quantitative and qualitative methods as well as a sound knowledge of social and policy analysis, market research, community development and customer care. This is difficult to achieve without either a reasonably sized research unit or a budget to hire consultants. It also means that chief officers need greater skills than in the past in marshalling and applying information from many sources.

Information flows are increasingly important to all public sector agencies. They organise their work and plan actions based on information feedback from the environment in which they operate. A research perspective brings a more structured approach to this information. This has been described as an 'action research' model:

> a data based, problem-solving model that replicates the steps involved in the scientific method of enquiry. Three processes are involved in action research: data collection, feedback of the data to the clients, and action planning based on the data. Action research is both an approach to problem solving – a model or a paradigm – and a problem solving process – a series of activities or events.
>
> (French and Bell, 1984, quoted in Cooke, 1992: 158–9)

An example of this model in practice is recent research on the link between housing and health. Rehousing on medical grounds has its origins in concerns with controlling infectious diseases, particularly tuberculosis, by relieving overcrowding. Tuberculosis is now much less common due to improvements in housing conditions and antibiotics. Today, medical grounds for rehousing are more likely to be related to long-term problems of chronic illness and physical disability. This is a consequence of the increasing proportion of older people in the population and the survival of more people with disabilities due to recent advances in medicine and medical technology (Cole and Farries, 1986).

It has been assumed for some time that rehousing people with medical conditions to better housing that is less damp, easier to heat, less crowded or better in some other way will result in a health gain. However, the time spent by GPs advising housing authorities on medical priority for individual applicants is a significant health service cost. In addition, rehousing on the basis of this advice appears to be neither well-targeted nor fair, as the following research findings indicate:

1. A survey by Cole and Farries (1986) of people rehoused on medical grounds in Bolton found that only 23 per cent thought their medical condition had improved, although a further 23 per cent were satisfied with the move despite their medical condition remaining the same. Only 14 per cent were still dissatisfied, mostly because the accommodation was not suitable to their physical disability.
2. A survey by Smith *et al.* (1993) of people rehoused on medical grounds in three urban authorities found that over 60 per cent of those rehoused experienced health improvements. But due to a shortage of suitable good housing, the policy of rehousing people on medical grounds was compromised by excess demand over supply. Only a minority of applicants secured priority rehousing, despite those denied medical priority often experiencing their illness as acutely as those awarded it. Two-thirds of non-movers believed their health had suffered as a result.

Cole and Farries (1986) conclude that a more realistic objective of rehousing should be to alleviate disability rather than to expect a medical improvement. They suggest that it would be better and cheaper to modify the person's existing accommodation to suit their capability, rather than to rehouse. Smith *et al.* (1993) also suggest that services and adaptations *in situ* may often be more appropriate. Both studies argue that rehousing remains the best option when it makes life easier for the tenant and their relatives by, for example, giving better access to close relatives and support networks.

Smith *et al.* (1993) propose that the policy of rehousing people on medical grounds should have its purpose redefined from cure to care. They argue that it is not possible to use medical rehousing as a cure for illness when (a) for many, there is unlikely to be a health gain, and (b) there is insufficient stock available in which to rehouse people with medical needs. However, some authorities are seeking to find ways of using their housing resources to bring about improvements in the health of council tenants, who now predominantly comprise the poorest sections of an authority's local population. Newcastle City Council is an example (Blackman, Harrington and Keenan, 1993). It has ended the use of GP assessments and, with health authority advice, introduced a more systematic assessment of health problems using a questionnaire either completed and returned by the applicant or administered during a visit by a housing officer. One in five medical priorities is sampled by a health visitor seconded from the district health authority, who also helps with the more difficult cases. Information from these assessments has the potential to enable better prioritisation of health needs in making decisions about whether or where to rehouse an applicant. For example, the health needs of families with children are now receiving greater attention. In order to explore the relative benefits of different actions, the Housing Department is carrying out further assessments of tenants following both rehousing and expenditure on housing improvements. In this way, the Department hopes that it will be able to build up a picture of what types of health problem benefit best from different housing allocation and investment decisions. This will enable the Department to make better-informed decisions about priorities for the use of its housing stock and the housing budget.

It is still the case that few public sector organisations could be described as having a 'research culture' of the type that integrates action-research into its operations. Decisions are often founded on informal theory or implicit beliefs held by professional officers or politicians. Research methods do not fit in very well to decision-making processes which are usually short-term and pragmatic, with decisions generally only involving marginal change from the current *status quo*. However, business planning and performance monitoring are developing rapidly as management practices in the public sector, and this is encouraging the clear statement of objectives and standards, and the identification of means of measuring achievement. This has created an environment more conducive to research.

In the National Health Service there has been a planned expansion of non-clinical research during the 1990s, focusing on needs assessment, health outcomes, management and consulting the public. In 1988, the House of Lords Select Committee on Science and Technology criticised the lack of connection between medical research and its contribution to health services. This led to the production of a national research and development strategy in 1991 (Department of Health, 1991). There has been a diversification away from biomedical and clinical research towards social science and social epidemiology, responding to growing evidence about the causes of ill-health and the differing cost-effectiveness of treatments and programmes (see Chapter 10). Managers and clinicians are being encouraged to seek and use the results of good, relevant research. Decision-makers are being urged to identify areas where research-based knowledge will be most likely to benefit them. Regional research and development plans have been developed, and Directors of Research appointed. There is a very large budget behind the strategy. The aim is to move towards spending 1.5 per cent of total NHS resources on research and development within five years.

Local government expenditure on research falls considerably short of this scale of commitment. The main reason for this is that research in local government is not recognised in the standard spending assessments used by central government to allocate revenue support grant to local authorities. It is a discretionary activity which must be resourced from the money allocated to running services.

There is a danger that the current review of local government being undertaken by the Local Government Commission and the Scottish and Welsh Offices will lead to a reduction in local government research. This is because the larger county councils are likely to be replaced by smaller unitary authorities. Although small authorities do conduct research by commissioning consultants, the loss of in-house research officers familiar with the policies and services of the local authority would be a major setback at a time when the need for policy-relevant research is growing.

NEEDS ASSESSMENT

The implications of the purchaser–provider split for research are particularly clear in the case of community care. Needs assessment is of major importance because the main justification for state funding and provision of social care is that this is the only way services can be provided according to need. Resource allocation should therefore be *needs-led* and not *demand-led*.

From 1 April 1993, state funding to support someone in a residential or nursing home depends on whether the local social services department assesses the person as having a need for this type of provision. Until this date, residential and nursing home provision was demand-led, funded by the Department of Social Security if the elderly person was in receipt of Income Support. The escalating cost of this provision led central government to give local authorities a 'gate-keeping' role, with cash limited budgets to purchase places in residential and nursing homes or to provide services in the community.

The need to develop policy responses to the ageing of urban populations is common to all the OECD countries of Western Europe, North America, Japan, Australia and New Zealand. A recent OECD report states:

> The ageing of the populations in OECD countries poses a challenge to urban governments. In the years of the post-war baby boom, governments strived to make cities good places for families to raise children. In the coming decades, local governments must ensure that cities are good places to age.
>
> (Organisation for Economic Co-operation and Development, 1992: 9)

Although older people who need care rely predominantly on informal sources of help, principally female relatives, the demand for public services from local authorities and health authorities will grow. The number of older persons living alone is rising. There is a rapid increase in numbers of very old people taking place. In the UK and other OECD countries, policy and programme responses are seeking to reduce institutionalisation in hospitals and residential homes, and to support ageing in the community by providing appropriate domiciliary services such as home helps and day centres. In the UK, all these services are financed through cash-limited budgets.

Planning for the changing age structure of urban populations has to be

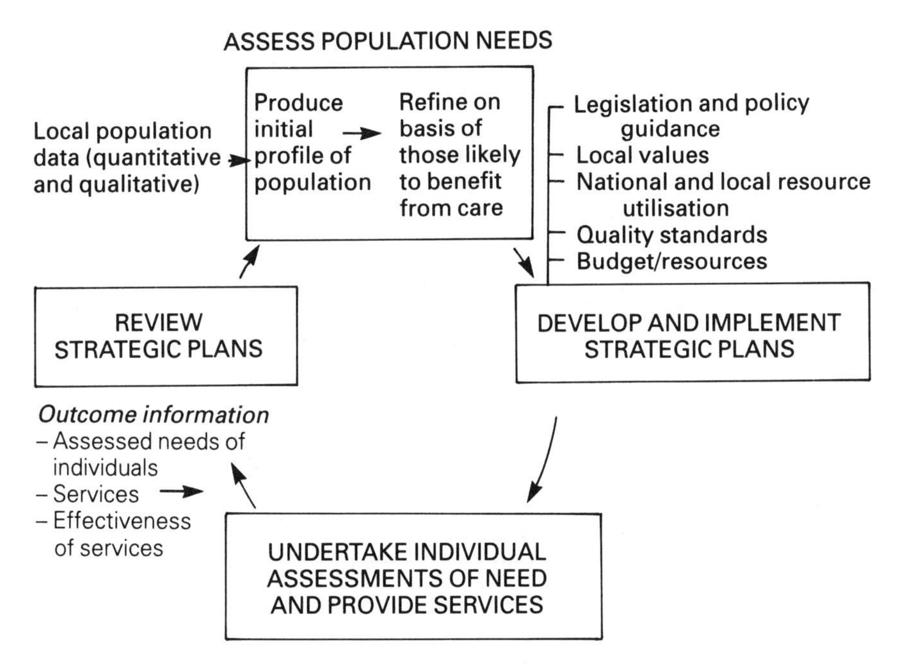

Figure 7.1 A population needs assessment model for community care
Source: Department of Health, 1993a.

based on two main activities. First, projections from current information. Second, analysis of the differences between projections and actual needs presented so that the current pattern of care services can be evaluated and options for change (including new services) identified.

The strategic and enabling role envisaged for local authorities in the NHS and Community Care Act requires that they carry out population needs assessments to estimate the numbers of people who fall into different categories of need and plan accordingly. This is because budgets have to be allocated to meeting different types of need, such as people with physical disabilities or people with learning difficulties, and estimates of the number of people likely to benefit from care have to be made. In addition, the provision of services in the statutory, voluntary and private sectors has to be managed and influenced so that resources are used to best effect. Figure 7.1 shows how the Department of Health envisages a population needs assessment model working in a local authority's social services department.

The first step in the Department of Health's model is to build up a demographic profile of the local population. Producing such a profile from primary research using a population survey is expensive, and there are alternatives. The national population census, carried out every ten years, provides information about the population structure at small-area level. At sub-district level, however, the data are not updated between censuses and so go out of date. Although this is often ignored, other methods have been developed with varying degrees of accuracy, such as annual enhanced electoral registration population surveys or use of Family Health Service Authority age/sex registers (Worrall and Rao, 1991; Dobson, 1992).

The Department of Health suggests estimating the prevalence in the population of particular needs, such as people with physical disabilities, by applying national prevalence data to local population figures (Department of Health, 1993a). A major source of prevalence data is the national disability surveys carried by the Office of Population Censuses and Surveys (OPCS) in 1985. These include, for example, estimates of the number of adults with different physical disabilities per 1,000 of the population, broken down into different age categories. Table 7.1 shows how the prevalence of need can be estimated in this way. However, users of these data are dependent on the definitions of disability used, which have been criticised (Miller and Munn-Giddings, 1993).

OPCS data also give prevalence rates for severity categories. This information can be used to group estimates into people with low, moderate and high needs. High need might be defined as people who need physical assistance every day; moderate need as people who need regular physical assistance because of mobility, communication problems or sensory impairment; and low need as people who need help intermittently. Such categorisation is necessary as social services departments have to prioritise the allocation of their limited resources based on levels of risk to people if services are not provided.

Great care has to be taken in applying national prevalence data to local populations. Factors such as local trends in the demand for and take-up of services, and the local circumstances in which people live, such as deprivation levels and the proportion of older people living on their own, will introduce variation between areas in the prevalence of need. An example of such a local factor is the higher than average level of disability found in old industrial areas where many older people have had their health affected by working in heavy industry. This makes estimates based on national averages inaccurate to a greater or lesser degree.

There are two main sources of *local* information against which estimates from national data can be compared. The first is administrative records containing details of people who have been referred to social services, assessed for their needs and are receiving services. These records are often computerised and held in a client information system. This is *expressed need* and may underestimate actual need for the following reasons:

1 Limited expectations of eligibility for assistance.
2 Lack of knowledge about potential sources of assistance.
3 Unwillingness to seek assistance from social services.

Table 7.1 Estimating prevalence rates for disability types

(a) District population figures

Age	*District estimate Year 1*	*District projection Year 2*
60–74	1,524 people	1,791 people
75 plus	468 people	690 people

(b) Prevalence rates per thousand population applied to above population figures

	Ages 60–74			*Ages 75+*		
Type of disability	*Prev. rate per 1,000 pop'n*	*Est. nos with disability in 1992*	*Est. nos with disability in 1993*	*Prev. rate per 1,000 pop'n*	*Est. nos with disability in 1992*	*Est. nos with disability in 1993*
Locomotion (mobility)	198	302	355	496	232	342
Reaching and stretching	54	83	97	149	70	103
Dexterity	78	119	140	199	93	137
Seeing	56	86	100	262	123	181
Hearing	110	168	197	328	153	226
Personal care	99	152	177	313	146	216

Source: Department of Health, 1993a.

4 Services are culturally inappropriate.
5 Needs are met in other ways, e.g. informal support or the involvement of another agency.

Social services departments should investigate hidden needs. For example, if a lot of need is met by family carers looking after older dependants, it may be necessary to target the development of relief and respite arrangements to prevent these informal care arrangements causing stress or breaking down.

The second main source of local information is surveys. For example, a survey could be carried out of a random sample of people over the age of 75, asking questions about their personal care needs and support networks. The sampling frame for such a survey could be the electoral register or the age/sex register of a family health services authority.

Local information from these sources may suggest that estimates from national prevalence data are inaccurate. If this is the case, it may be necessary to adjust projections from national prevalence rates to take account of local factors.

The next step in the Department of Health's modelling of needs for social care is to apply 'preferred' or 'typical' care packages to people with a given type and level of need. Such packages should be the subject of consultation with users, carers and care workers. A typical package of care for an older person assessed to have a medium level of need for social care might be three hours per week home care, three days a week at a day centre and six trips a week transport to the day centre. By costing these packages and multiplying them by estimates of the number of people in different need categories social services departments can estimate the provision required. If this exceeds the budget available, further prioritisation has to be carried out by raising the threshold of need necessary to qualify for services or amending the care packages to reduce their cost.

The main purpose of projecting need is to plan expenditure by basing budgets on actual types and level of need, to stimulate the right services and to encourage complementary action by other agencies such as community health services and voluntary organisations. These projections can be compared with expressed need recorded through individual needs-based assessments of people referred to social services. The assessment forms should be designed so that information can be categorised in a way consistent with the population profile.

Table 7.2 shows an example of aggregated information from individual assessments for a given year compared with the projections for that year. In this example, there are large variances between the local authority's projections of need and the actual numbers of older people with a moderate level of need seeking assistance. The variances will need to be further researched. Three possible reasons for them are:

Table 7.2 Comparing projected need with assessed need

Needs of elderly people	*Original projections for 1993*	*Nos. seeking assistance for this disability*
Locomotion (mobility)	697	402
Reaching and stretching	200	110
Dexterity	277	105
Seeing	50	43
Hearing	423	210
Personal care	520	450

Source: Department of Health, 1993a.

1 The projections were based on very little factual information about informal care, resulting in less need than anticipated having to be met by formal services.
2 This district has a large minority ethnic population with lower take-up of services than might be expected due to cultural barriers.
3 In some areas, such as hearing problems, the social services department has never publicised the services that are available.

Similarly, information on actual care packages provided can be aggregated to compare actual expenditure on services with projected expenditure. Table 7.3 shows for illustrative purposes an example where, although the number of older people receiving services was lower than expected, the care packages they received cost the local authority almost as much as it had expected to spend on a larger group of people.

A further stage in this type of approach is for desired outcomes for care packages to be stated. Actual outcomes can then be compared with desired outcomes. For example, desired outcomes over a given period of time for an older person living alone might be to *improve* their mobility level (with aids) from a score of 4 to a score of 2 and to *maintain* their ability/skill level in bathing themselves at a score of 3, as against a predicted deterioration without intervention to a score of 5. This information about outcomes can then be recorded and aggregated as shown in Table 7.4. Although monitoring outcomes in this way will only give indications about the effectiveness of different service strategies and care packages, it is likely to highlight issues for further investigation.

The example was given above of using national prevalence data with local demographic data to estimate the incidence of disability in a local population. Demographic and other census data are important indicators of need and are preferable to information about demand or activity as means of deciding on the allocation of resources. This is because, as long as a good relationship exists between census variables and the needs upon which services are targeted, census variables are unaffected by such factors as differences in the

Table 7.3 Comparing projected with actual expenditure

Need group: elderly people with moderate level of need
Projected population: 160
Actual number of care packages: 102

Total provision provided:		
	408 hours per week home care	£2,448
	306 days a week of day centre	£6,120
	612 transport trips per week	£1,224
	204 meals per week	£612
	30 occupational therapist visits per week	£300
Cost of provision	*£10,794 per week*	

Source: Department of Health, 1993a.

Table 7.4 Comparing desired with actual outcomes

Need group: elderly people
Level of need: moderate
Period: October to December 1993

Task area	*Number of care packages including task area*	*Number of improvement desired outcomes achieved*	*Number of maintenance desired outcomes achieved*
Mobility (with aids)	30	15	10
Bathing	27	19	4
Food preparation	11	6	3
Continence	15	6	7

Source: Department of Health, 1993a.

level of service or awareness about services among the public. These factors can affect demand and activity, making them an inaccurate reflection of actual needs.

Until 1991, Newcastle City Council allocated resources for providing home care workers to its six area social services teams on the basis of historic expenditure, or activity. Thus, each area team's allocation depended on what it spent the previous year. In 1991, the social services department decided to establish a formula from census data to enable resources for the home care service to be deployed across the six area teams in a way that would relate to indicators of need rather than purely to historic expenditure.

The 1992/3 formula for home care incorporated data from a 1986 city-wide survey of residents carried out by the City Council. The indicators used in such formulae should be tested out against survey evidence or the views

Table 7.5 Newcastle City Council's formula for allocating the budget for home care to six area teams

Indicators of need	*Weighting multiplier*
1 Overall population in area	×1
2 Number of 85+ people in the area	×6
3 Number of 75+ people in the area	×3
4 Number of people with long-term sickness in the area	×3
5 Number of people 65+ in the area	×1
6 Number of people in non-white ethnic groups in the area	×0.5
Deprivation indicators weighted at 0.8 per indicator (5 × 0.8 = 4)	×4
1 Non-car ownership in the area	
2 Number of people who moved house in the last year	
3 Lone-parent households	
4 Number of unemployed people	
5 Number of local authority tenants	

Source: Newcastle City Council Social Services Department, 1993.

Table 7.6 Gains and losses resulting from Newcastle's new formula for allocating home care and resources

Area	*Share of budget (%)*	*Actual budget*	*Gains and losses in home care worker posts*
1	11.951	£612,264	no change
2	17.669	£905,163	– 15 posts
3	22.665	£1,161,096	– 2 posts
4	11.907	£609,990	– 5 posts
5	15.497	£793,902	– 4 posts
6	20.308	£1,040,337	+ 9 posts

Source: Newcastle City Council Social Services Department, 1993.

of users and service providers. In Newcastle, consultations were carried out with social work managers and home care staff to decide on (a) what variables should be included in the formula and (b) what weight each variable should have, given that some variables and weightings would be more strongly associated with their home care workloads than others. In deciding on the variables and weightings, account was also taken of historic expenditure.

In 1993/4, Newcastle's total home care budget was £5.3 million. The formula based on Table 7.5 was used to arrive at a percentage distribution for each of the six social services areas in the city, apportioning the total budget accordingly. The 1993/4 formula, which used 1991 census variables, included more up-to-date information than the 1986 city-wide household survey upon which the formula for 1992/3 had been based, and incorporated the new census variable: number of people with a long-term illness or disability. Population change since 1986 meant that the new formula had a sharp effect on home care posts in the area teams (see Table 7.6). In view of this, the social services department decided to phase in the reallocation over two years.

Need analysis in public services is not new but has acquired a higher profile in recent years. Although the above examples relate to community care, the principles are similar for other public services where the following concerns apply:

1 Assessments of need should be based on what is best for the individual rather than on the individual's suitability for existing services. Assessors should therefore not be providers.
2 Unmet need should be recorded and future need identified.
3 Priority groups should be targeted, with a particular focus on vulnerability and risk.
4 Hidden need should be identified.
5 Outcomes for service users should be defined.

None of this is as simple as it might seem. Needs-led services are nevertheless constrained by the resources available. Unmet need is virtually infinite and may not be a need for new services but for more of the same. In practice, public services have to establish useful indicators of need, adopt eligibility criteria for different service levels within their resource constraints, and monitor the level of need compared with the level of service provision. Research skills are particularly useful in these areas. For example, Gibbons (1991) compared the amount of social services received by families in different priority need groupings. She also monitored the outcome of service provision over a four-month period, using scores on various indicators and parents' ratings of whether their problems had improved. Her conclusions illustrate how this type of research can point to where improvements are needed:

> [R]eferred families with the greatest needs were most likely to receive some help from the social services department. However, families with lesser, but still significant need levels were no more likely to get any help than were families with the least needs. Improvement in social workers' assessments at the point of referral should lead to even better matching of services to needs. This is important, since families with higher levels of need appeared to benefit most from social services' practical help.
>
> (Gibbons, 1991: 225)

Gibbons found the provision of day care to be particularly effective in terms of bringing about a measurable improvement in families' problems.

Many practitioners, however, will point out that assessing an individual's needs is not as straightforward as completing a questionnaire in a single interview. The process can take time. People may have low expectations or not want to be dependent even though they are at risk. Miller and Munn-Giddings (1993) list the following factors which can prevent a 'need' becoming expressed:

1 The person is not aware of their 'need' or does not consider it to be serious enough to warrant seeking assistance.
2 From what the person knows about existing services, he or she does not believe that something can realistically be done to alleviate the need.
3 The need is not felt to be a priority on which it is worth spending their time and possibly money compared with something else.
4 The person has help from family or friends or the money to buy help privately.
5 The person does not know that some appropriate service or assistance is available.
6 The service that is available is not perceived in a positive light, for example, it is stigmatised or would involve too great a loss of privacy.
7 The costs for the person or a carer, such as travel or inconvenience, are felt to outweigh the likely benefits.
8 The person does not know where to make contact with a relevant agency or is not able to do so.
9 The person, or someone on their behalf, does not have the skills to make out the existence of the 'need' effectively.

Once a need is expressed, public sector provision depends on whether the need is judged to be eligible for a service. Eligibility may not result in the delivery of a service straight away but involve a waiting time, with priority according to the level of need. The availability of a service will depend on the amount of supply.

Need assessment draws on people's views but this still only informs and complements existing perceptions and policies. It is a type of consultation but not a full partnership. Percy-Smith (1992) argues that 'social audits' are an alternative which try to go beyond traditional needs assessment to involve local people in assessing their own needs. She draws on Doyal and Gough's (1991) definition of well-being as, 'the ability of people to participate in life'. A social audit involves the following stages:

1 Collation of all available information about the target area or group, including demographic, health and economic data, and information about existing services. This information is compared with city, regional and national figures.
2 A survey of the local population to investigate their own views of their needs, combining the survey with focused discussion groups with different sections of the community.

3 Information about needs as perceived by professionals working in the area, using interviews.
4 Collation of this information into a draft report which highlights issues where there is agreement and issues where there is little consensus. Discussion of the report at public meetings involving statutory and voluntary agencies, user groups, community organisations and members of the public. The purpose of these meetings is to increase understanding of the issues and to find agreement about the most important needs, how they should be met, by whom and at what level.
5 A final report with recommendations. This acts as a benchmark against which achievements and failures can be assessed.

Percy-Smith (1992: 34) points out that, 'need does not entail rights ... it does not necessarily follow that simply because a set of needs has been articulated there is any absolute duty on the part of policy makers to allocate resources in such a way that those needs are met. In an environment of scarce resources hard choices will still have to be made.' Her approach in fact raises many of the same issues as community development, which seeks to go beyond needs assessment to establish community involvement and to build on capacities within the community (see Chapter 6).

Resource allocation can be informed by research about needs, but it is another question whether resources allocated on the basis of need actually achieve the benefit intended. This is especially the case with regard to policies of positive discrimination towards geographical areas which have high rates of deprivation ('priority areas'). Census data are widely used to define such areas, but these data are only available for geographical areas rather than individuals (with the exception of the samples of anonymised individual and household records available from the 1991 census, which are too small for most resource allocation purposes). As a result, the use of a geographical framework for mapping deprivation and allocating resources is fairly common. It is reinforced by the fact that public services are often organised territorially into administrative areas and that local politics is based on the political territories of electoral wards.

Territorial needs are not simply the sum of the individual needs of the population in the area. The attributes of the locality itself can influence needs for public services and the concentration of deprived groups in some areas may tend to produce a 'multiplicative effect of deprivation' (Curtis, 1989: 141). Macintyre, MacIver and Sooman (1993) identify the following elements of areas which can disadvantage people independently of their personal circumstances:

1 Physical features of the environment shared by all residents in an area, such as air and water quality.
2 The availability of healthy/unhealthy environments at home, at work and at play. For example, healthy foodstuffs are less available locally, and

often more expensive, in deprived areas. Access to healthy recreation such as sporting facilities is often not as good as in better-off areas.

3 Services provided, privately or publicly, to support people in their daily lives. Car ownership is low in deprived areas but there is usually no compensatory better provision of public transport services. The quantity of primary health care services such as GP practices, dentists and opticians is often lower.
4 Socio-cultural features of a neighbourhood. Deprived areas often have a higher incidence of problems such as vandalism, litter, muggings, assaults, disturbances in the street and burglaries. There may not be the local leadership or resources to campaign for better services.
5 The reputation of a neighbourhood. People living in deprived areas can find themselves discriminated against in applying for jobs and credit.

This would appear to justify area-based policies in relation, for example, to housing, health services, transport, policing, recreation and retail food provision. However, although targeting measures on particular areas can be an efficient way of delivering resources, it misses people with the same problems who may live outside these areas and its effectiveness within targeted areas depends on how well the measures reach individual members of their target groups. As well as researching the distribution of needs, it is important to research the processes which produce these needs and the factors which may affect access to services (Curtis, 1989).

Recognising the existence of 'area effects' separate from individual deprivation is one part of this. The features of local social and physical environments are important focuses for urban policy and can be changed by local action. However, it is important to retain a focus on what happens to individuals even when the policy is area-based, otherwise unintended consequences are more likely. An example is displacement caused by neighbourhood renewal. These programmes often diversify the housing tenure of an area, converting or building properties for owner-occupation in neighbourhoods previously almost wholly comprised of council housing. Bradford and Steward (1988) used census data for enumeration districts to evaluate whether displaced council tenants moved into 'better' or 'worse' areas following the transfer of their housing to private developers for refurbishment. There was an improvement for many tenants, but some ended up in worse housing.

Aggregate geographical data can be useful in evaluating area effects and targeting resources. But they can be misleading as surrogates for individual data, leading policy-makers to the wrong conclusions. Classifications of areas based on geographically aggregated data label areas according to their relative concentrations of target groups, e.g. high proportions of people who are unemployed or households with children under 5 years old. These still often form only a minority of all residents. There is a tendency, however, for

Table 7.7 Correlations between educational attainment, single-parent families and deprivation in one region of Scotland in 1985

Correlation between educational attainment and the following:		
Level of data aggregation	*Lone-parent families*	*Deprivation index*
School	−0.40	−0.89
Neighbourhood	−0.11	−0.43
Individual	−0.06	−0.31

Source: Marsh *et al.*, 1991.
Note: Educational attainment is a 14-category variable scoring grades attained at Scottish Certificate of Education O-grade and Higher examinations; the deprivation index is an additive score formed from 12 census indicators at enumeration district level selected by a factor analysis and weighted according to factor scores; the data are from the Scottish Young People's Survey 1985 and the 1981 census.

all people living in the area to become labelled by these generalised stereotypes. This is called the 'ecological fallacy' and can also give rise to some statistical issues.

Table 7.7 shows negative correlations between (a) the educational achievement of children in the Scottish Young People's Survey and (b) living in a deprived area or coming from a lone-parent family. In row one the data are aggregated at school level. In row two, the data are the same but aggregated at the level of the various neighbourhoods served by the schools. Both correlations are weaker at this level of aggregation. The bottom row shows the results of correlating the data for individuals. The correlation between exam success and coming from a lone-parent family becomes negligible, and that between exam success and living in a deprived neighbourhood becomes much weaker (Marsh *et al.*, 1991).

QUALITATIVE RESEARCH

Quantitative research is necessary to establish answers to 'how many' and 'how much' questions. But understanding why things happen is better investigated with qualitative methods which collect data about perceptions, attitudes and behaviour (Blackman, 1993a).

Vittles (1991) discusses the results of a survey by York City Council which showed that 50 per cent of respondents were in favour of a new covered market in the city centre and 50 per cent were in favour of an open market. Surveys such as this which ask people straight yes/no or agree/disagree questions are common ways of gauging public opinion but do not answer 'why' questions in any depth. Decision-makers will often want the answer to 'why' questions, especially when quantitative survey results are inconclusive and fail to reveal the strength of feeling behind people's answers. York followed up this survey with qualitative research using focused discussion groups. This found that those in favour of an open market held their views much more strongly, and for clear environmental reasons. Those in favour of the covered market were really motivated by a desire not to get wet and their views were very weakly in support of the covered option.

Although more public sector organisations are now undertaking or commissioning qualitative studies of their services, most research in the public sector is strongly quantitative. Its concerns have traditionally been with analysis of secondary datasets such as the population census, and with surveys. Quantitative data are often preferred because they appear 'objective' and prove things 'statistically'. Qualitative data are perceived to be subjective and unreliable. Warren (1991) explains this as follows:

> qualitative research is under-used by government, partly because its essentially 'soft' nature makes it appear less authoritative than its quantitative cousin, and secondly – more specifically – because policy-makers will naturally tend to demand research that generates 'facts', since there is then less of a gulf between the research and the decision-making that must follow.
>
> (Warren, 1991: 13)

'Facts' in this context are 'objective' pieces of information, preferably numbers and statistics. However, no 'facts' can exist independently of the presuppositions imposed on them. Quantitative data are interpretations produced by researchers, coders, government and business. Concepts such as 'health', 'ethnic group' or 'unemployment' are open to interpretation and different definitions.

Qualitative data can in fact be more valid than quantitative data because people's responses to questions are not fitted into the pre-determined categories of a questionnaire but analysed as the respondent's own words. Hedges and Kelly (1992), for example, undertook a qualitative study of community identity for the Department of the Environment to inform guidance to the Local Government Commission in its review of local government structure. Their findings were based on twenty focus groups in a mixture of different types of area in different parts of England. By using focus groups, these researchers were able to develop an understanding of how people identify with different kinds and levels of community in different contexts. They were able to distinguish between 'clan' communities, identification with 'home ground', and 'communities' of social contacts. A questionnaire survey would have been inappropriate for investigating such an ill-defined topic.

Eight circumstances can be identified when it is particularly appropriate to use qualitative policy-related research:

1 When the interest is to provide background and explore attitudes (why do black people under-utilise certain services?).
2 When a greater depth of understanding is needed (how can a service be improved for those who express dissatisfaction?).
3 When motivation and processes that people do not easily articulate need to be examined (attitudes to HIV/AIDS or satisfaction with services among people with mental illness).
4 When little is known about the subject (how does the poverty trap affect household decisions about employment?).
5 When the target audience knows little about the subject (reporting research on racism to white people).
6 When the issues are high profile with 'positions' being taken (policy on sexuality).
7 When the subject to be researched is 'dull' and needs to be made interesting (quality of refuse collection service).
8 When the issues are highly complex (where should budget reductions be made?).

User satisfaction is often investigated better using qualitative rather than quantitative research. This is especially the case in social services where results from snapshot quantitative surveys can be misleading. Miller and Munn-Giddings (1993: 51–2) list the following reasons for this:

1 Respondents tend to express a high degree of satisfaction unless they can think of something in particular they want changed. They may think that a negative response could mean a reduction or withdrawal of their service.
2 Respondents are unlikely to know much about alternative services.
3 Respondents often evaluate their social workers rather than the care services they receive.
4 Respondents may be 'captive clients', such as parents with their child taken into care, and therefore unable to express choice.

An advantage of qualitative methods is that they enable researchers to explore why people feel the way they do and to explore alternatives. Working with groups can reveal issues that may not emerge from individual interviews. Miller and Munn-Giddings give the following example:

> Groups are important in facilitating the exchange of ideas and experiences and have the potential as an independent lobby. At a recent feedback session on some consumer research in Essex with agencies and carers who had taken part in the 'respite care' study, we were struck by the suggestions that arose as a result of one carer voicing an opinion with which others then concurred on an issue that had not arisen in interviews.
>
> (Miller and Munn-Giddings, 1993: 51)

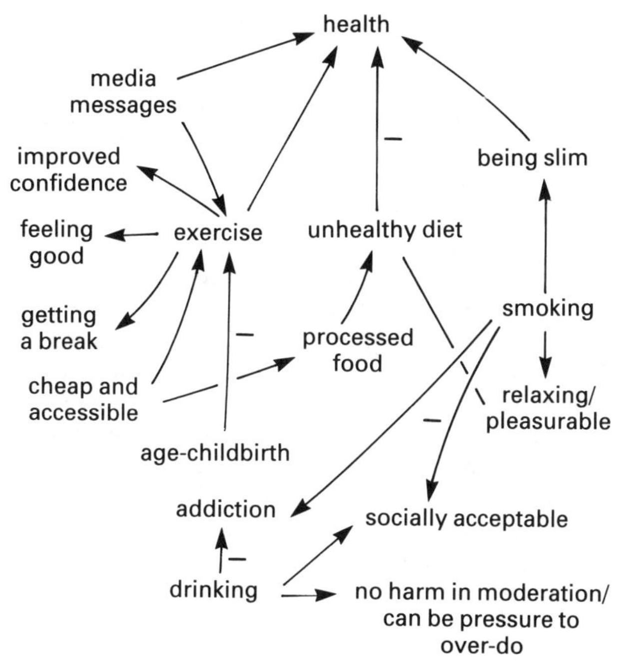

Figure 7.2 Health among women sports centre users in Newcastle upon Tyne
Source: Blackman, 1993a.

Qualitative research is about studying local and small-scale social situations. Its methods involve close and sustained encounters with people. They include focus groups, in-depth interviews, participant observation and techniques such as 'mystery customers' who test how easy it is to get in touch with the right person and how the user is treated. Qualitative data are verbal/ conversational or documentary. They are difficult to analyse. Analysis has to look for the structure of what people are saying, often using coding rules. Cognitive mapping has been used to link respondents' expressed ideas together with the implication that these are relationships of 'may lead to' or 'affects'. Figure 7.2 shows an example from an analysis of a focus group about health among women sports centre users in Newcastle. Among the findings of this research were that media messages are influential in promoting health and exercise, and that improved confidence, 'feeling good' and getting a break were among the benefits that the sports centre provided for these women.

Qualitative research is difficult with more than a small number of respondents and therefore cannot establish prevalence and ranges in the same way that quantitative research can by attaching percentages to the results. However, it can be corroborated by other qualitative studies, and indeed Dryzek (1990) argues that qualitative case studies carried out in some numbers can form the basis for judgements about the typicality of their findings and is preferable to survey research. Dryzek questions whether the quantitative research approach of averaging across a large number of observations and deriving law-like regularities from the data is appropriate at all in investigating public opinion. This view is echoed in Titterton's (1992) criticism of need assessments which work with crude aggregate categories such as 'elderly people', rather than recognise and cater for individual differences (see Chapter 2).

Much local government research seeks to establish what people need, what conditions exist (such as local demography and the incidence of deprivation) and what change has occurred. Managers are generally interested in research to answer 'how many', 'how much' and 'what is the evidence on' questions. This biases research designs towards studies that produce figures upon which generalisations can be based. There is a danger that this approach reproduces a power relationship between those providing and those receiving services. People using services have to be involved in the early stages of decision-making: in deciding how to assess the performance of services for example. Runciman (1983) argues that social science seeks to reduce the possibility of misapprehension. He suggests six ways this can arise, and all are very relevant to urban policy:

1 *Incompleteness.* Are policies achieving what is expected?; are service users receiving everything they should?; are needs being met as intended?
2 *Oversimplification.* Are performance indicators measuring actual performance and comparing like with like?
3 *Ahistoricity.* Are people's needs understood in the context of their history and milieu? Redevelopment in the 1960s, for example, often destroyed communities and social networks.
4 *Suppression.* Are the views of all the 'stakeholders' in an issue represented? A survey of public opinion about housebuilding in Newcastle's greenbelt, for example, found that over 50 per cent of employed respondents opposed releasing greenbelt land for housing, but among unemployed people only 28 per cent opposed the proposal (Newcastle City Council, 1992b).
5 *Exaggeration.* Controversial issues such as youth crime are prone to exaggeration. Policy responses must be informed by research. Although juvenile crime is actually falling, the number of offences is increasing as a result of an increase in the number of persistent offenders upon which preventive measures now need to be targeted.
6 *Ethnocentricity.* The needs of minority ethnic people need to be researched directly and not imputed from what is known about the needs of white people. Important differences exist between ethnic groups within the black population.

No single research design is appropriate to investigating these issues. Often a combination of quantitative and qualitative research is most appropriate: qualitative to improve understanding and quantitative to establish extent and correlations.

CONCLUSIONS

Urban policy needs to be informed by research about (a) the scope of services and needs in localities; (b) what issues and changes in services should be focused on; and (c) what objectives, targets and performance measures should be used. Once an issue is identified, it needs to be formulated as a research problem, setting out the questions to which answers are wanted. This will involve considering one or more of the following research designs:

1. *Qualitative research*. This is appropriate when a greater depth of understanding is needed (such as reasons for dissatisfaction or under-use of services), or when the issues are sensitive, little is known about them or they are highly complex (such as where to make budget reductions). Methods include focus groups and in-depth interviews. Although allowing issues to be observed and explored in depth, qualitative research will not normally establish how typical the findings are.
2. *Social surveys*. These use standardised questionnaires and samples. They enable representative generalisations to be made from the findings but often lack depth. Obtaining large samples which enable results to be analysed by sub-categories such as age groups and local neighbourhood problems to be picked up is expensive. Postal questionnaires, although economical, can only successfully substitute for interview surveys when the questionnaire is short or highly relevant to the respondent.
3. *Social indicators*. The population census is the prime source of secondary data about social conditions. Indicators can also be derived from administrative records, such as social services and housing assessments. Another type of social indicator is user complaints.

The use of research has to be facilitated organisationally. There should be regular meetings between researchers, managers and policy-makers about the research agenda for the organisation. Networks of advice and support should be developed and used, involving, for example, local academics as advisers and research collaborators.

Research should be subject to the same processes of performance monitoring as other services. This applies particularly to the dissemination and implementation of research findings. Nuyens's (1991: viii) comments about health research apply across all policy-related research:

> [if research] is to be accepted as a managerial tool or an instrument for more informed decision-making, then it should be made clear why and

> how it makes a difference, which developments followed the research, which decisions were taken on the basis of the research findings, which process of social change was initiated and/or facilitated through the research, and why all this happened . . . or did not happen.
>
> (Quoted in Hunter and Long, 1993: 45)

Often research does not appear to be practically useful. It refines the definition of problems and provides partial solutions but can leave policy-makers with more questions than they first started with. Policy-makers therefore need to have a better understanding of how to use research. There are two main aspects to this.

First, it is unrealistic to expect any single piece of research to answer the growing complexity of problems which local government and other organisations are attempting to address. Huxley *et al.* make the following points:

> First, the results are usually less salient by the time the research is complete. Second, all researchers have 'angles' and all studies have defects. Third, policy problems usually involve more variables than are used in the research.
>
> (Huxley *et al.*, 1990: 187)

Second, the role of research in a *political* decision-making environment can be unclear. It is easy to despair when facts and figures are presented to council committees only to be overridden by political calculations. Ritchie's (1993) account of using research to support a group of parents opposing a school closure is an example:

> Despite the range of information and analysis and scale of objections raised, the schools committee of the County Council took the decision to close the school. The reasons given were precisely those which the group had challenged, including straight factual errors about the type of construction of the school. It was clear that all the evidence produced had been ignored.
>
> Overwhelmingly the message from this involvement is that the information which the community needed related to how decisions are really made (the political process) and how to access that process.
>
> (Ritchie, 1993: 89)

It is quite wrong to expect that research is anything more than a tool used for particular purposes. It rarely provides definitive answers. It is not value-free. Research therefore cannot substitute for decision-making based on political and professional criteria.

Research is also not a substitute for consultation with the public, although it does have a special role in producing valid and reliable information about issues (Blackman, 1992).

Urban policy is never likely to be 'research-led' because of the extent of executive and political discretion that is involved. Research has an advisory role. Reade suggests that:

> What researchers and advisers should . . . be primarily concerned with, I would suggest, is analysis of the feasibility of political objectives in technical terms, as well as the clarification of them, and with advising on the means of their achievement.
>
> (Reade, 1987: 199)

There is a danger, however, that the proximity of research to management and political processes will bias the activity away from studies that might produce uncomfortable evidence or be seen to challenge current policy. The 1990 National Health Service and Community Care Act is a good recent example. This legislation introduced an approach to policy-making and service delivery based on assessing need. The government seemed to be encouraging local authorities to research need and to consult users and carers, formulating their strategies for meeting these needs jointly with district health authorities and family health services authorities, and publishing them as annual community care plans. The main consequence of introducing needs-based approaches is that unmet needs start to be measured. This is uncomfortable for governments seeking to restrain public spending.

The House of Commons Health Committee reported on community care following its investigation of the area during the 1992/3 session. The Committee made detailed recommendations about monitoring the adequacy of government assumptions about required resources. The government's response to these recommendations has been to backtrack on earlier policy advice to monitor unmet need and to argue that this is in fact not possible because defining need is too difficult. Previous guidance to local authorities to aggregate up the results of individual assessments of need and to feed this into the community care planning process was retracted:

> The Committee suggests that a national picture of need for services could be calculated from the sum of individual assessments and community care plans. The Government does not share this view. The assessment of needs and decisions about services are distinct processes. Trying to identify and measure need solely on the basis of front line workers' views of the services they would like to provide to individual clients, regardless of competing claims on resources, would be potentially subjective and divorced from reality.
>
> (Departments of Health and Social Security, 1993: 4)

Policy-makers will only be advised by research, not led by it. Percy-Smith's comment about social auditing is true of research in all areas of public policy:

> what it cannot do is provide a ready-made formula for making the hard

> decisions in a technocratic manner (e.g. whose needs should be met and at what level?) Choices will still have to be made and priorities set. Those are necessarily *political* choices and are properly made by fully accountable politicians.
>
> (Percy-Smith, 1992: 31)

Research can therefore help politicians to make better choices based on fuller information. Indeed, the same information might also provide the public with benchmarks against which the achievements and failures of politicians can be assessed.

Part III

KEY GOALS FOR URBAN POLICY

8

EDUCATION, TRAINING AND THE LOCAL ECONOMY

Two key goals of urban policy are to stimulate the economic development of depressed areas and to target training and employment measures on people who are disadvantaged in local labour markets. The National Commission on Education (1993) emphasised the importance of raising general standards of education and training if the UK is to improve, or even maintain, its economic position in the world economy. The Commission identified as an urgent task tackling disadvantages caused by deprivation, gender, race, disability and geography which prevent large numbers of people realising their potential at school, in employment and in terms of personal and social development.

This chapter considers education, training and local economic development together. It begins by identifying the nature of the UK's economic difficulties and then focuses on the tasks of improving and monitoring the level of education and training. The range of work of local authorities is described, including local skills audits, supporting community businesses and making economic development plans for their districts, together with arguments for them to have a strategic role in the local economy. The chapter concludes with a review of the impact of the European Union on local economies and their resources for economic development.

EDUCATION AND TRAINING

In the early 1900s Britain was still in a very powerful trading position despite having lost its place as the world's leading industrial nation. Textiles, clothing, iron, steel and coal were exported in large quantities. These industries relied heavily on Third World markets and the existence of the British Empire. During the years of world economic depression between 1918 and 1939, demand stagnated and traditional export industries declined sharply. There was, however, some diversification into new sectors such as chemicals, vehicles and electrical engineering. After the Second World War, a long period of world economic growth saw the economy recover. But this expansion masked continuing weaknesses. With the reappearance of global recession in the mid-1970s, British industries fared badly in competition with other countries. Manufacturing industries were particularly affected and a series of redundancies and closures ensued. By 1991, living standards in the UK ranked eighteenth among the 24 member countries of the Organisation for Economic Cooperation and Development, and eighth among the twelve countries of the European Union (National Commission on Education, 1993).

A range of factors is involved in the UK's present poor economic position. These include:

1. Under-investment in manufacturing industry.
2. Slow adaptation to new methods such as team working, multi-skilling, flexibility and total quality programmes.
3. Low wages in manufacturing industry.
4. Weak national strategies for education, training and investment.
5. Weak support to small businesses and their roles in innovation and as flexible and responsive suppliers to large firms.

There is growing recognition in Britain that tackling these problems will necessitate giving a higher priority to knowledge and skills. The recent report of the National Commission on Education (1993) comments:

> In an era of world-wide competition and low-cost global communications, no country like ours will be able to maintain its standard of living, let alone improve it, on the basis of cheap labour and low-tech products and services. There will be too many millions of workers and too many employers in too many countries who will be able and willing to do that kind of work fully as well as we or people in any other developed country could do it – and at a fraction of the cost. There are several examples today of insurance companies which have their clerical back-up work done in countries where education is adequate and labour is cheap. Computer software is already often written in countries far distant from where it is to be used. These are pointers to the future.
>
> (National Commission on Education, 1993: 33)

The UK's future as an advanced industrial economy will depend on high-skill and high-technology industry. Investment in people is the key to competitiveness in these sectors; the Confederation of British Industry estimates that by the end of the century at least 70 per cent of all jobs will need a professional level of skills (Lowe, 1992). Broadly speaking, this means more technicians and fewer operatives, and more managers and fewer supervisors.

Trends towards 'knowledge jobs' have been evident for some time. Between 1971 and 1990, the number of managerial, professional and technical jobs in the UK grew by 3.1 million, while the number of skilled manual jobs declined by 1.1 million and the number of jobs for machine operatives and other less skilled occupations fell by 1.8 million (Wilson, 1993).

The UK has not been particularly successful in applying its scientific and technical knowledge base to producing new products and services that can compete internationally. Government policies towards science, technology and engineering were recently reviewed, leading to publication in May 1993 of a White Paper. This announced important changes to the organisation of state funding for research in science, engineering and technology (Cm 2250, 1993). These changes seek to link research in universities and other government-funded research centres more closely to the needs of industry and other users of research, including the public sector.

Whilst such a strategy is able to build on a good science and technology base in UK universities, there remain serious shortcomings in the scientific and technological skills of the British workforce. These originate in too few young people obtaining good qualifications at school. Only 27 per cent of 16-year-olds in England obtained grades in the range A to C in GCSE mathematics, English and one science in 1990/1. Achievement at an equivalent level in Germany was 62 per cent, France 66 per cent and Japan 50 per cent (Green and Steedman, 1993). The proportions of young people obtaining a comparable upper secondary school qualification at age 18 plus in 1990 were 29 per cent in England but 68 per cent in Germany, 48 per cent in France and 80 per cent in Japan. Similar differences exist for vocational qualifications.

The UK's educational provision for 16- to 19-year-olds and the provision of training for young people in general has not been good. Until the establishment of training and enterprise councils (TECs), government training initiatives were largely focused on relief programmes for unemployed people such as Employment Training (ET) and Youth Training. One effect of this has been to isolate and devalue training, which has widely been seen as a way of keeping unemployment figures down. For example, a recent survey of attitudes among British and German young people found that British young people saw training schemes as a poor alternative to a job and placed a low value on vocational qualifications (Lowe, 1992). Training for people already in employment was also neglected until TECs opened this up by providing 'training credits' to help pay for courses for suitable applicants.

In 1991, the government published a two-volume White Paper, *Education and Training for the 21st Century*, which sought to increase the participation of 16- and 17-year-olds in education and training and raise their level of attainment (Employment Department Group, 1991). The new policy, implemented by the 1992 Further and Higher Education Act, means that every 16–17-year-old is given a 'training credit' which entitles them to buy education or training from a college or employer. The policy also brings within a single diploma framework both the academic-related qualifications of 'A' and 'AS' levels and work-related qualifications such as BTEC National

Table 8.1 The NVQ framework

	Qualifications of different awarding bodies (e.g. BTEC/SCOTVEC) accredited by the National Council for Vocational Qualifications		
NVQ level	*General NVQ (broad occupational areas)*	*Occupationally-specific NVQ*	*Equivalent academic-related qualification*
1 Competence in a range of mostly routine activities	Pre-vocational certificate	Semi-skilled	National curriculum
2 Competence in a range of activities, some complex and non-routine, with some individual responsibility	Broad-based Craftsmen Foundation	Basic craft certificate	GCSE
3 Competence in mostly complex and non-routine activities in a wide variety of contexts with considerable responsibility	Vocationally-related National Diploma. Advanced Craft Preparation	Technician. Advanced Craft. Supervisor	A/AS levels
4 Competence in a range of complex, technical or professional activities, often with responsibility for others and for resource allocation	Vocationally-related Degrees/Higher National Diploma	Higher Technician. Junior Management	Degree
5 Competence in a range of complex techniques and principles in often unpredictable contexts, with very substantial responsibility	Vocationally-related Postgraduate Qualifications	Professional Qualification. Middle Management	Postgraduate degree

Source: Developed from Employment Department Group, 1991.

and Higher National Certificates and Diplomas, which are accredited by the National Council for Vocational Qualifications. National and Scottish Vocational Qualifications (NVQs and SVQs) are based on national standards of competency expected in employment rather than just knowledge, and can be obtained by accumulating credits for units over a flexible period of time (see Table 8.1). They are being introduced in all sectors of the economy in order to raise the skills level of the UK workforce. To date, some 80 per cent of occupations are covered by NVQs and SVQs at five levels of competence. The new framework is a move towards simplifying the UK's fragmented post-16 qualifications. But traditional 'A' levels were retained, despite extensive criticism that they are too academic, cover too narrow a range of subjects and only cater for a small minority of the population at post-16 level. The result is that the aim of trying to secure parity of esteem for academic and vocational studies is likely to be undermined.

Since the early 1980s there has been an upward trend in the proportion of 16- to 18-year-olds participating in education and training. However, nearly 90 per cent of those who will be in the workforce in the year 2000 have already left compulsory education. There is thus concern about the level of education and training of the existing workforce over and above that of young people leaving compulsory education. Poor achievement in education and training results in a higher risk of unemployment. The 1990 Labour Force Survey revealed that almost 11 per cent of economically active people of working age with no formal qualifications were unemployed, compared with less than 3 per cent for graduates and others with qualifications above GCE 'A' level. Forty-three per cent of unemployed people had no formal qualifications, compared with 25 per cent of all people of working age.

The greater number of workers in the UK with poor qualifications, or no qualifications at all, has contributed to social division. The National Commission on Education, in reviewing the disturbing failure of the UK to educate and train its young people effectively, warned that if this continues:

> Those whose education had enabled them to become 'knowledge workers' and who, as people owning the resource most vital for success in the marketplace, were able to command relatively high salaries and safe jobs would be in a position of remarkable power and privilege. Those denied such an education would be excluded, and the exclusion would become more painful with the passing of time as the less skilled jobs continued to disappear and pay in the jobs that were left fell further behind . . . The threat to social cohesion is obvious.
>
> (National Commission on Education, 1993: 37–8)

The widening of access routes into training and higher education is a positive development. Education and training are essential to the future of the UK, but, as Miliband writes, there are particular skills that are needed:

> All the evidence suggests that for Britain to maintain its current position in the second division of industrialised countries, we need a substantial increase both in our staying-on rates at 16, and in our output of educated people from higher education institutions.
>
> Today, it is the skills of abstraction, thinking in systems, and experimentation (as well as the thoroughly contemporary concept of working in teams) that [are] the central props to competitive advantage in a modern economy.
>
> (Miliband, 1992: 7)

It is particularly important that all qualifications let students acquire a core set of transferable skills as well as a grasp of a particular body of knowledge. These skills can be defined as:

1 Communication skills.
2 Numeracy.
3 Personal and interpersonal skills.
4 General information technology skills.
5 Problem solving skills.

These skills reflect the contemporary needs of both industry and the public sector. Essentially, these are:

1 Adaptability to change.
2 Taking a creative approach to tasks.
3 Employees taking responsibility for their own work and development.

In 1991, the Confederation of British Industry published a series of national education and training targets. These were developed jointly with the Trades Union Congress and other major partners in education and industry. They were subsequently adopted by central government and are important targets for upgrading education and training. Four of the targets are for 'foundation learning', as follows:

1 By 1997, 80 per cent of 16–24-year-olds should reach NVQ level 2 or the equivalent (i.e. GCSE with at least four passes at A–C level).
2 Training and education to NVQ level 3 or the equivalent (two 'A' levels) should be available to all 16–24-year-olds who can benefit.
3 By the year 2000, 50 per cent of 16–24-year-olds should reach NVQ level 3 or equivalent.
4 There should be education and training provision for young people to develop self-reliance, flexibility and breadth.

There are also four targets for 'lifetime learning' which apply to the workforce as a whole. These are:

1 By 1996, all employees should take part in training or development activities.

2 By 1996, 50 per cent of the workforce should be aiming for NVQ/SVQs or units towards them.
3 By 2000, 50 per cent of the workforce should be qualified to NVQ/SVQ level 3 or equivalent.
4 By 1996, 50 per cent of medium to large organisations should be 'Investors in People'.

Investors in People is a programme run by the Employment Department which encourages employers to invest in training. Funds are provided through TECs. Businesses which join the scheme have their current training and development practices reviewed. They then adopt an action plan which is followed by an assessment by external licensed assessors.

These targets require a dramatic increase in the number of people taking part in training. The target of 80 per cent of young people reaching NVQ level 2 by 1997 compares with a proportion of only 51 per cent in 1991 and an annual increase of only 1–2 per cent recently (National Commission on Education, 1993). The target of 50 per cent of young people reaching NVQ level 3 by the year 2000 compares with a proportion of only 30 per cent in 1991 and an annual increase of less than 1 per cent recently. Achievement of the lifetime learning targets requires a similarly dramatic increase from the present position. Better-qualified people entering the workforce from school or college will not be sufficient to meet the targets. There is therefore an urgent need to promote adult learning and training.

SCHOOLS

Recent years have seen many initiatives in schools to integrate work experience, economic awareness and enterprise into the curriculum. Schools are now offering BTEC and other NVQ qualifications. The careers service also has a very important role to play in providing individual advice, counselling and career planning to young people at the age of 16.

Central government initiatives have promoted technology teaching in schools. City Technology Colleges were launched in the 1980s as partnerships between business and central government, although only 15 have been established. Some 220 schools have received enhanced funding for technology teaching under the Technology Schools Initiative. In 1993, the government published proposals for Technology Colleges (Department for Education, 1993b). These will be existing schools which will have business or industry sponsors and have a strong emphasis on technology teaching. Additional government grants will be available. The schools, however, must be grant-maintained – that is, they must have opted out from the control of their local education authority.

Inner city schools were targeted by the 'compacts' initiative which has sought to increase employer involvement in schools and raise pupil performance. By the end of 1993/4 there were 62 compacts operating in the UK, involving over 800 schools and 10,000 employers. Young people join the compact in their final two years and make a personal agreement to fulfil goals such as regular school attendance, consistent punctuality, work experience and completion of school coursework. A survey in 1992 commissioned by the Employment Department found that improvements in pupil performance were 'small but measurable' (*Employment Gazette*, March 1994).

There are some areas that have been neglected in recent policy developments. Adult education is provided by local authorities through colleges and, most successfully in terms of access, through freestanding centres in the community. National education policy is to provide vocational courses through the independent further education sector, leaving local authorities with 'leisure' courses which it is assumed will be self-financing. Although local authorities subsidise these courses for people on benefits, it is likely that the charges necessary to achieve self-financing will close off an important route for many adults back into education and training.

The push for training has also lacked a strong emphasis on reaching people who find it difficult to learn. Such students are financially unattractive for colleges to recruit. Provision has to be made for people to learn at a suitable pace and level. This is a special case of the general need for education and training to be available in suitable modules and with credit transferability between the various pathways to qualifications.

The National Commission on Education (1993) made a series of recommendations to tackle weaknesses in the UK's education system. Among the most important of these are:

1 The phased introduction of high-quality nursery education for all 3- and 4-year-olds, starting with deprived areas. The Commission accepted evidence that nursery education is linked with later success at school, as well as enabling women to seek employment.
2 Rationalisation of the National Curriculum and the introduction of a new General Education Diploma to replace GCSEs, 'A' levels, BTEC and existing vocational qualifications. The Commission advocates a target for the year 2000 of 90 per cent of young people working for nationally recognised educational qualifications at least until they are 18 years old. Adults should be entitled to off-the-job education and training.
3 A series of measures to raise the overall quality of the teaching profession, encourage and evaluate innovation in teaching, raise literacy and numeracy levels, reduce primary school classes to a maximum of 20, and ensure that all pupils receive continuing guidance throughout their schooling.
4 Integration of the management of education and training at central

government and local level, with the creation of education and training boards to replace local education authorities.

The Commission estimates that its recommended measures would cost an additional £3.2 billion a year, although this would reduce to £1.4 billion a year by the end of the century if its proposals for students to contribute to the cost of higher education courses are included.

Schooling should enable children and young people to fulfil themselves in a variety of ways. It has a wider purpose than just economic. Education policy should encourage a positive attitude towards learning in young people, with primary schooling in particular having a crucial role in giving all children confidence in their ability to learn and succeed whatever their particular capabilities. Song, drama, sports and practical skills for instance are all valid and important routes through which children express and fulfil themselves. However, education policy must not lose sight of the importance of equipping people with the language, reasoning and number skills needed to absorb new information and adapt to change. It is unlikely that a school, for example, will be effective in delivering wider benefits to pupils if it is failing in the central objective of pupils acquiring skills and knowledge.

Educational achievement should be a major goal of urban policy. The means to achieve this goal include introducing more effective teaching and learning methods, and providing more resources in areas where clear gains have been demonstrated, such as nursery education and reducing primary school class sizes. In addition, measures are needed to reduce the barriers to achievement caused by inequalities arising from income, gender, ethnic group, area of residence and the effectiveness of the school or college. Considerable evidence is now available demonstrating the marked effects of both teaching methods and resources on achievement, as well as effects caused by factors such as deprivation and variation between schools and colleges in the quality of their leadership, expectations of pupils and parental involvement. Much is already known about how to optimise achievement and reduce the effects of these arbitrary factors (National Commission on Education, 1993).

Urban policy has an important role in targeting resources in order to equalise opportunities. In education this includes planning the provision of nursery education and allocating resources for special needs in schools. By monitoring the effects of such measures the best use of resources can be determined and the most successful practices identified and shared between schools. Local education authorities monitor school examination results, but these are of little use as indicators of effectiveness. Although schools are required to publish their examination results, these do not measure school effectiveness because the prior attainments of schools' intakes strongly influence how well their pupils subsequently perform in exams.

Comparisons of examination results between schools and subjects within schools must therefore take into account differences in prior attainment. A number of local education authorities use 'value-added' approaches to do this. These identify the boost or 'added value' that a school or subject within a school gives to a pupil's previous level of attainment. This is done by measuring the difference between an earlier and a later attainment in a test or exam, i.e. the pupil's *progress*.

Measuring value added entails a longitudinal analysis of many individual pupil performances. One method of presenting the results is to group pupils by GCSE score into low, middle and high achievement. This can be done for all pupils in, say, the local education authority and for pupils in individual schools. The average 'A' level score in each of the three GCSE groups can then be calculated for the LEA pupils as a whole and for each individual school. Schools which score above the LEA average in a group may be adding more value to their pupils' GCSE results and schools which score below less. However, comparisons must take into account variability between individual pupils with the same prior attainment; there will be a range of achievement within the same group, and this range will differ between schools. In other words, there will be a margin of error associated with each school's average. Comparisons based on value-added measures should therefore use broad-brush statements such as 'well above' expectation or 'as expected' rather than finely ordered rankings.

Whole school summaries tend not to be favoured. Subject-by-subject analyses give more useful and practical information (McPherson, 1992). Monitoring over time should also be used, rather than taking a single snapshot of one year's results, because schools' performances change from year to year. In addition, it is known that the characteristics of the pupil body as a whole are correlated with the progress of individual pupils. Some studies have shown that a pupil of average prior attainment or from a deprived home background will make more progress in a school where the prior attainment of all pupils is above average than in a school where the prior attainment of all pupils is below average. Analysis should take such 'peer-group effects' into account.

More sophisticated approaches to value-added use multi-level modelling techniques to differentiate between factors such as gender or age in measuring added value. These can demonstrate whether, other things being equal, schools add different levels of value for different types of pupil. For example, one school might give a special boost to pupils who are already doing well, another might achieve the greatest learning gains for pupils who have previously struggled, and a third might succeed in boosting the science attainment of girls in particular. Thomas, Goldstein and Nuttall (1993) found that schools had significantly different value-added scores at 'A' level for pupils who achieved in the low, middle or high range at GCSE, describing this as 'differential effectiveness'.

Although value-added approaches ideally use pupils' individual test or examination results as input and output data, input measures of prior attainment may not be available. In the absence of such measures it is possible to use several other items of information about each pupil's background, such as socio-economic circumstances. Even if exam results are not available, tests can be specially administered to provide schools and colleges with baseline data. These can also provide information of diagnostic value in providing teachers with guidance about areas where an individual student may need additional support.

Value-added information can be used to target where improvements should be expected. The main focus of attention should be on schools and subjects that do significantly better or worse than expected, establishing what others schools and subjects can learn from them. Extra support and advice should be targeted to bring about improvement in areas which perform below expectation.

To date, work on value-added approaches has been largely confined to schools. It has been extended with some success to the further education sector. Methods are being developed to measure the value added by vocational programmes, although these present technical difficulties because of the diverse attainment at intake and the less differentiated grades that are awarded for achievement (FEU, 1993). The Audit Commission has extended the use of value-added approaches in school effectiveness studies to compare school performance with schools' costs per pupil in a value-for-money framework.

Linking knowledge about school effectiveness to strategies for school improvement is a developing field. Indicators of school effectiveness should be regarded in the same way as performance indicators; they point to where there may be exceptionally poor or good performance but do not measure quality directly (see Chapter 5). There may be several possible reasons why performance appears to vary.

LOCAL ECONOMIC DEVELOPMENT

Education has long been a local government responsibility, although the role of local education authorities is being eroded by the local management of schools and other changes. Economic development, however, expanded as a local government activity during the 1980s. This was in part a response to a policy vacuum at central government level; economic difficulties concentrate geographically yet there was no single agency responsible for making a corporate response to these problems at local level (Hasluck and Duffy, 1992). Although budgets and staffing levels are very small compared with mainstream services, economic development often has a high political priority in local authorities and is an area of policy which has corporate significance.

The last decade or so has seen a mushrooming of economic strategies in local government. But in fact local authorities have a long history of involvement in the economy. At the turn of the century, direct ownership of municipal enterprises was common. Glasgow was the most 'municipalised' city in Britain, with council ownership extending across housing, tramcars, gas, hospitals, police, water and markets. These early municipal economic initiatives had the overall aim of providing the public with better and cheaper local services than could be provided by private enterprise.

After the Second World War, local authorities were a prime vehicle for creating the welfare state. Although their services expanded, their role in economic development declined. However, in Scotland, some of the English regions and Northern Ireland a role continued in regional planning which combined regional industrial policy with physical planning and urban renewal (Danson and Lloyd, 1992). Regional planning was particularly important in the Glasgow, Edinburgh and Belfast metropolitan regions in the 1950s and 1960s. It appeared to hold considerable promise for the following reasons:

1 It could take into account the broader context or process within which local economic problems occur, such as technology limitations caused by a lack of training provision in a region.
2 It could plan integrated packages on a large scale, such as new towns, together with industrial sites and new housing opportunities, allowing slum clearance in other areas.
3 It could integrate national policies with local policies based on regional priorities.
4 It could provide a framework for agencies to work together and implement policies more efficiently.

The promise of regional planning, however, was often not realised. The deindustrialisation of regions such as Northern Ireland and North-East England in the 1980s included much of the new investment attracted by regional planning measures in the 1960s and 1970s (Blackman, 1988). The failure of the economic base of the new towns and 'growth centres' created by regional planning caused high unemployment and a problem of vacant, difficult-to-let housing estates.

Regional planning also contributed to the decline of inner cities by promoting alternative greenfield sites for industry. This decline began to be recognised in the 1960s with the development and launch of the Urban Programme (see Chapter 3). This programme sought to give priority to deprived urban areas in public expenditure programmes. In the 1980s the emphasis of the programme was shifted to encouraging private enterprise, supporting projects that would have measurable benefits, and avoiding projects which in the long term would continue to demand public expenditure.

The Thatcher government elected in 1979 shifted national economic policy away from state intervention in industry, including winding up regional planning. Government spending was channelled through new agencies, principally UDCs, enterprise zones and TECs. These have had briefs which are local and single-purpose and are led by people with commercial and industrial backgrounds.

Until the 1980s, local government economic policy largely related to land-use planning and the provision of industrial sites and small-scale factory premises. Infrastructure provision, promotion of the area to investors and public relations were also undertaken. The beginning of the 1980s saw a dramatic worsening of conditions in many local economies. Deindustrialisation occurred as firms closed down or moved out of inner city areas. National economic policies were essentially allowing market forces to remove uncompetitive industries. In response, some local authorities worked with trade unions and community groups to oppose company closures, but these campaigns had little success. Policy shifted to strategies aimed at supporting positive restructuring of firms and industrial sectors.

A few of the larger urban local authorities established enterprise boards, such as the Greater London Enterprise Board and the West Midlands Enterprise Board. These carried out research and developed local economic strategies and initiatives. Selective investment from council funds was aimed at supporting sectors which had prospects for growth and diversifying away from sectors which were clearly in decline. New approaches were encouraged for traditional industries such as footwear and clothing, particularly *flexible specialisation* by small and medium enterprises which local authorities had the resources to assist.

Flexible specialisation involved supporting the introduction of new technology, such as computer-aided design and manufacture, to enable products to be manufactured in small batches rather than in mass production. Instead of producing standardised products for mass markets, flexible specialisation involves a high level of product variety targeted on precise population types or 'market segments'. Large companies decentralised design, innovation and production, presenting opportunities for local authorities to work with groups of small businesses, provide training and plan city centres and neighbourhood centres as places for both work and leisure.

The abolition by central government of the Greater London Council and the metropolitan county councils in 1986 largely brought to an end this phase of local authorities making selective investments on the basis of local sectoral policies (Geddes and Benington, 1992). Local councils have been encouraged by central government policies to work in public/private partnerships in which local government powers are used constructively to support company's needs for land, infrastructure, training and other factors. A key aim has been to attract inward investment, particularly Japanese and American firms seeking to position themselves with the Single European Market.

Local authorities thus continue to have an economic development role, but this is limited. It has to be pursued in partnership with government agencies such as the UDCs and TECs. And although many local authorities are giving greater attention to intervening in the local labour market with education and training initiatives, and to establishing links between schools and local businesses, their powers to undertake these initiatives have been reduced (see Chapter 3).

Hepworth (1990) lists the following factors which local authorities have most sought to promote or develop:

1 Cost and availability of land and property.
2 Cost and availability of labour.
3 Transport infrastructure.
4 Image of the area.
5 Telecommunications infrastructure.

Despite its fifth place, telecommunications are having a growing role in local economic development for the 'information city' (Hepworth, 1990). Some local authorities see cable television as a way of developing community-based media, distance learning and local information services, although this could be undermined by the dominance of commercial interests in cable. Yeomans (1993) paints a picture of the 'intelligent housing estate' with telecom links bringing tenants distance learning opportunities, employment information and improved access to services. Of particular importance to peripheral regions and cities in the North of England, Wales, Scotland and Northern Ireland is the possibility of using advanced telecommunications infrastructure to overcome geographical marginalisation and promote these areas as efficient locations for companies. In the South of England there is a different interest in using telecommunications to reduce commuting and pressure on central London.

Local economic policy has also been important in maximising job creation from activities which are not primarily economic development. A good example is housing investment. This is an important economic tool; about 30,000 jobs are created or saved for every £1 billion spent on housing investment. Redevelopment and improvement programmes in high unemployment areas are now frequently linked into construction training programmes targeted on local people. In addition, houses built for sale in depressed council estates can bring in people with buying power to support local shops and businesses.

Local government economic policy cannot have a marked effect on local economies, but at the margins it can deliver tangible and cost-effective benefits. Its success will always primarily depend on the strength of the local economy. However, irrespective of economic conditions, Hayton (1989) found that a few key factors were generally responsible for the success of those projects which do get people into jobs. The two key ones are:

1 Carrying out research prior to establishing the project. This needs to focus on the labour needs of local employers. Such information can be gathered quite simply from scanning job advertisements in the newspapers and personal contacts with employment agencies. Training has to be appropriate to the jobs available. Some jobs may be highly skilled and need technical training to industry standards, such as computing or electronics. Many jobs, however, are low skilled, such as in the hotel and catering sector. Employers are looking for people with the right personal aptitude and attitude, making the confidence and motivation of job applicants particularly important.
2 Recognition that the long-term unemployed, who are the main target group for local government economic policy, are likely to have a variety of needs other than skill training. These are often linked to confidence and motivation. They need to be addressed with personal counselling and support.

Apparently successful projects can be affected by problems of *displacement* and *low pay*. Displacement occurs when getting unemployed people into jobs has the effect of displacing others in the labour market who are in equal need. Low-paid and low-skilled employment is unlikely to be a force for the regeneration of local economies. Lasting impacts which minimise displacement and low pay depend on raising the skills of residents in high unemployment and deprived areas so that better-paid employment becomes a possibility. Raising awareness of career opportunities and providing follow-up training for those who do enter relatively low-skilled work are important in seeking to achieve this objective.

Local councils, however, now have a minor role in training. This is largely limited to their education departments and schools, supporting prevocational training and school-based initiatives. However, many authorities work closely with local TECs to identify training needs. Their officers are often in day-to-day contact with employers, potential inward investors, unemployed and self-employed people. Skills audits may be funded by local authorities (see pp. 214–15). Prevocational training is particularly important in deprived areas, with a focus on raising awareness and building confidence among people who have been unemployed for long periods.

Local authorities have long supported local businesses by providing grants, advice and training. Many councils continue to employ officers who work in partnership with TECs and other agencies to identify training needs, run small business courses and provide customised training targeted on unemployed people who want to set up community businesses such as launderettes, food co-ops or child-care schemes. Although small-scale, this work can have significant benefits. Some examples of the type of support that is offered are:

1 *Counselling support and financial assistance* to new and 'pre-start'

businesses, long-term unemployed people and 'community enterprises'. This support is often decentralised through structures such as local employment development agencies in areas with high unemployment.

2 *Helping local businesses to expand* by identifying premises for expansion or relocation, assisting with loan and grant applications, advising on training, and accessing relevant services from other business support organisations.
3 *Providing low-cost accommodation* in subsidised office, workshop and studio units. Some local authorities provide this accommodation in business support centres with communal workshop facilities, business services, security and on-site management and business advice.
4 *Helping businesses compete in Europe* by providing information and advice about EU regulations, legislation and standards. Local authorities are also a source of advice about EU grants, training initiatives, and research and development support.
5 *Providing financial assistance* through loans and grants for business start-ups, equipment purchases, market research, premises improvements and property rental.
6 *'One-stop' shops* putting business enquirers in touch with other council services such as property information, environmental protection or planning services.
7 *Help with complying with regulations and legislation.* Many business activities require licences, planning permission or other forms of compliance with legislation. Local authorities are responsible for many of these areas, including planning permission, building regulation, environmental health, enforcing the Shops Act, street trading, trading standards, and regulating private sector providers of child care and homes and services for elderly and disabled people.

In recent years there has been a more critical attitude to the role of support to small businesses in creating jobs in areas of high unemployment. McArthur (1987) found that conventional small firms in the Glasgow area employed very few people who had been unemployed for longer than a short period. Conventional small firm development was unlikely to offer much to long-term unemployed people.

One response to this has been to support 'community enterprise'. This involves local residents in high unemployment areas in initiating, managing and controlling *community businesses*. The effects of economic recession in the 1980s stimulated many community and voluntary groups to try to develop projects which would create jobs. Community businesses emerged as small trading organisations owned and controlled by local people who join as members, usually by paying a small membership fee, and in turn elect members to form a volunteer board of directors. Local authorities often supported this community-based approach. This was especially the case in

Table 8.2 Recruitment to community enterprises and conventional small businesses in Glasgow

	Community enterprise (%)	*Conventional small business (%)*
Previously employed	14	31
Unemployed less than a year	9	62
Unemployed a year or more	69	7
Other	8	1
Live in local area	76	42
Live elsewhere	24	57

Source: Donnison *et al.*, 1991.

urban Scotland where the number of community businesses grew from only a handful in the late 1970s to an estimated 140 in 1990, with a reported annual turnover of some £13 million (McArthur, 1993).

Community businesses have been supported financially by Urban Programme funding to start up local shops, catering and cleaning services, repair businesses, security services, light engineering businesses and other mostly low-tech labour intensive trading organisations. This has been successful, on a very modest scale, in targeting unemployed people and in providing public services in areas that were poorly served in this respect (see Table 8.2). However, very few of these projects have been able to establish commercial viability. Among the factors which appear to be associated with community business failures are not adequately monitoring financial performance and the lack of business capabilities among volunteer directors. McArthur concludes, from a study of a community business failure in Glasgow, that:

> There are convincing arguments in favour of giving special treatment to community businesses because they operate in economically difficult localities and are trying to employ workers who have been excluded from the labour market for long periods. However, the ready availability of grants and contracts, without introducing other checks and supports geared towards the community business's organisational capacity, can carry a huge risk ... The importance of finding a good manager or management team cannot be overestimated [and] specialist secondees in areas such as accountancy, marketing, financial management and business development are the sorts of skills which could be usefully added.
>
> (McArthur 1993: 227–9)

McArthur suggests supporting community businesses through contracts with their funding agencies. The contract would specify the composition of the board, the regularity of meetings, the relationship between the board and management, the production of financial monitoring information, and training requirements.

A variant of the community business is the credit union. Credit unions are essentially financial savings and loans co-operatives owned and run by their members. Credit union members must share a common bond, usually by living in the same locality or working for the same employer. Between 1988 and 1992 the number of credit unions affiliated to the Association of British Credit Unions increased from 84 to 236, the number of credit union members from 30,034 to 73,089, and assets from £9 million to over £29 million (Association of Metropolitan Authorities, 1993b).

Credit unions are important in a number of ways. They help people to save and borrow on a modest scale, offering some protection against the problem of extortionate and often illegal money lending in deprived areas. They can serve as a catalyst for the personal development of their members, who run the credit union co-operatively. And they can assist with community development by providing a useful service, encouraging participative democracy and sometimes funding community activities from their profits.

Skills audits are an economic development tool fairly widely used by local authorities and TECs to establish information about local labour markets. This information is collected to plan local intervention on the demand side of the labour market: the skills of the existing local workforce. The audits usually take the form of a survey of existing skills among the local adult population living in a specific geographical area, generally one of high unemployment. They are sometimes complemented by a survey of firms to achieve a picture of the demand side for skills. Skills audits are the only way to obtain detailed and up-to-date data on the local labour supply. Local authority economic development units have used them to identify mismatches between the type of jobs available and skills needed, and the skills which exist. This informs work on which skills need to be created through targeted training and re-training initiatives to improve job prospects in high unemployment areas.

Employers' views on skills shortages need to be interpreted carefully. Haughton and Peck's (1988: 13) experience is that, 'many employers . . . claim skill shortages when really the underlying issue is that they are underpaying or they are exaggerating their own experience, in part reporting local hearsay'. These authors cite the results of other research which suggests that:

> the key shortage was not of those with the appropriate conceptual and motor skills but of the *right kind of people* with those skills. That is, personal factors such as a potential employee's perceived attitudes to trade unions heavily influenced employers in their responses.
>
> (Haughton and Peck, 1988: 14)

A particular focus of many skills audits is to identify the barriers which result in certain groups of people, such as women or black people, being excluded from full access to jobs and training opportunities.

Skills audits can use postal questionnaires to firms and local residents. A much better response rate is achieved by using face-to-face interviews, although this is more expensive. Some studies link training into the survey by specially training local unemployed people to conduct the survey.

A further development of the skills audit is to use the exercise to set up a local register of people who are available for work, with a listing of the skills they possess. When job opportunities arise, such as the construction of a new shopping centre, this information can be sent to people with relevant skills who are on the register.

Haughton (1991) summarises typical labour market barriers which recent local skills audits have revealed. These are as follows:

1 A need for retraining.
2 A lack of local information and guidance on training.
3 Low pay and training allowances.
4 A lack of childcare provision.
5 Travel distances and times.
6 Health problems.

He also lists the policy recommendations typically made by skills audits:

1 Improve vocational education and training.
2 Improve information about self-employment.
3 Improve information available to people seeking training and jobs and to employers.
4 Improve childcare facilities.
5 Improve local transport availability.

Skills audits frequently discover very high turnovers in labour markets in areas of high unemployment. People move in and out of the area, in and out of employment and unemployment, and between different employers. Although getting someone into a job may appear to be a successful outcome, a common labour market experience for unemployed people is short-term and casual work. In addition, people who are successfully assisted by a training initiative are also very likely to move out of the area if it is unpopular. One way of counteracting this is to link training initiatives to improving the local environment, making areas of high unemployment more attractive places to live and thus also possibly improving attractiveness for employers as well as employees.

Skills auditing has proved to be a useful tool and is probably sufficiently important to be a continuing monitoring process rather than be undertaken as one-off exercises.

A very significant aspect of local councils' involvement in their local economies is their considerable *purchasing power.* Although central government legislation constrains the extent to which local councils can implement policies such as equal opportunity objectives through their purchasing from private sector companies, many councils still seek to achieve policy objectives through their contracts with the firms that supply them with goods and services. For example, councils can require firms wishing to be considered for council work, or under contract to the council, to comply with equal opportunities and health and safety legislation. They can evaluate and monitor firms against criteria set out in legislation and codes of practice.

The Local Government and Housing Act 1989 makes the preparation of *economic development plans* a statutory duty for local authorities. These are annual documents which describe the programme to be carried out in the coming year, together with output targets and expenditure. Local authorities are required to consult about the content of the plans, a provision which is aimed at getting the business community more involved with local authorities (Fenton, 1992). The local council, however, determines the form the plan takes. This has ranged from costed lists of activities to very detailed strategies. The best examples describe the local economic context and set out proposed initiatives, explaining the reasons for them, when and how they will be implemented, and the criteria for judging their success or failure (Thomas, 1992). Economic development plans also address strategic issues such as the impact of the Single European Market, poverty and environmental sustainability.

In the London Borough of Hammersmith and Fulham, where an economic development strategy was first produced for the borough in 1978, the annual economic development plans have an important role in raising awareness among the authority's own councillors and chief officers about the role of local economic development. Fenton (1992: 22) describes this as explaining, 'what we are trying to do, how we are going to do it and what the benefits to the people of the borough will be'. The borough's economic development unit develops projects in the following areas:

1 Direct service delivery, e.g. business advice and grants.
2 Setting up, and support to, borough training and business organisations.
3 Coordination of council corporate economic development initiatives.
4 Joint initiatives with the private sector.
5 Lobbying to change government schemes, e.g. through the Association of London Authorities.
6 Identifying new key issues for policy.
7 Commissioning research and feasibility studies.
8 Securing external resources.

Local economic regeneration is marked by fragmentation of responsibility. Whilst the economic development plans primarily state the local authority's proposals, there are other key players in local economic development. These include chambers of commerce, UDCs, TECs, enterprise agencies and government task forces. Thomas suggests that economic development plans are used to establish the extent of overlap in local economic development services and to identify new initiatives where gaps exist:

> Local authorities have an important role to play in setting the overall strategy and in influencing and coordinating the activities of the other local actors. For example, Sheffield has an ESCHER strategy which not only incorporates an overarching structure for the local authority's own separate policies – covering environment, social, cultural, health and economic regeneration – but has also secured the agreement of the TEC, UDC and the private sector that they will pursue their own activities within this wider strategic framework.
>
> (Thomas, 1992: 118)

Fenton (1992) makes a similar point in explaining how the London Borough of Hammersmith and Fulham's economic development plan has the purpose of looking at where present gaps are and prioritising target groups, sectors of industry and types of project.

THE SINGLE EUROPEAN MARKET

The Single European Market, completed at the end of 1992, will have significant effects on local economies. Although intended to stimulate economic growth, increased competition could severely damage weaker local economies. Urban policy will have to respond by strengthening local planning, transportation, environmental and training policies. However, more than a local response is needed. A recent Coopers and Lybrand/ Business in the Community report, *Growing Business in the UK*, argues that powerful regional development bodies need to be established if the UK is to be competitive in Europe (see Chapter 11).

The Audit Commission (1991b) has urged local authorities to assess the effects of the Single Market and to review their organisational ability to meet new requirements and exploit new opportunities. It identifies three main ways in which the European Union is affecting local government:

1. *Euro regulation*: implementing, enforcing and monitoring EU legislation.
2. *European economic integration*: identifying new opportunities for the local economy.
3. *Eurofunds*: potential support for the local economy and/or council projects.

It is important that local authorities develop strategies to respond to these factors, based on a review of needs. Clear responsibility for aspects of the strategy needs to be allocated within the authority, without 'ghettoising' European affairs in the council's structure. Information flows from EU institutions need to be managed, and this may need specialised skills. Local authorities can act as 'Euro-facilitators' for the local economy, such as helping local employers and organisations gain access to information about grants, contracts advertised within the EU, and new European standards. They also have an important lobbying role, seeking to influence draft directives and to support grant applications.

European liaison officers, and in some cases teams, are now common in British local government. Strathclyde, Birmingham, Greater Manchester, Cornwall, Kent, Lancashire and the East Midlands have local authority staff located in Brussels, whilst other authorities have organised representation through consultants or joint arrangements. However, the Audit Commission has warned that care has to be taken to achieve value for money, suggesting that for every ECU an authority spends lobbying in Brussels it should win ten to spend locally. It advises that individual local authorities are unlikely to find locating staff in Brussels cost-effective compared with the selective use of direct contacts, visits and *ad hoc* alliances. Bongers (1992) also stresses the need for local authorities to co-ordinate their lobbying, especially over policy issues. The Council of European Municipalities and Regions (CEMR) has an important role in this respect, acting as the representative body for the national local government associations, and has built up a strong presence in Brussels. The Local Government International Bureau is also having an increasingly important role in providing UK local authorities with intelligence about EU regulatory and legislative developments.

In the area of local economic policy, many local authorities are already working with local employers and chambers of commerce to assess the threats and opportunities posed by the Single European Market. This includes such work as developing possible trade connections or identifying firms vulnerable to European competition because of reliance on the home market, particularly in the public sector. A number of local authorities and local development agencies run pilot local employment projects sponsored by the EU and aimed at strengthening small and medium-sized firms. A number of local authorities are also involved in the SPRINT programme, which provides part-funding for facilitating the sharing of technology and industrial innovations. Other programmes are industry-led, such as the EUREKA programme which is aimed at strengthening European technological capability.

EU legislation on public contracts for works, supplies and services requires that larger contracts let by central government, local authorities and other public sector bodies are advertised throughout the European Union and do not discriminate against continental tenderers. Although there have been relatively few cross-border tenders to date, these are likely to increase. The legislation includes services such as maintenance and repairs, construction-related work, refuse disposal, computing, accounting and auditing which are, or will shortly be, subject to compulsory competitive tendering in the UK.

An objective of the Single Market is to ensure that services and products will be freely marketable in all member states. To facilitate this, uniform European technical and trading standards are replacing different national standards. There is a large research and development programme supporting the development of common standards. Local authority contracts and enforcement work have to comply with these standards. As employers, local authorities also have to comply with harmonised EU health and safety requirements and the mutual recognition of qualifications (although the British government's rejection of the Social Chapter of the Maastricht Treaty means that European standards relating to employment conditions and worker participation will not be implemented in the UK).

The introduction of the Single European Market was accompanied by a doubling of the EU's 'structural funds' to tackle regional economic disparities within the Community. These are the European Regional Development Fund (ERDF), the European Social Fund (ESF) and the Guidance Section of the European Agricultural Guidance and Guarantee Fund (EAGGF). Between 1994 and 1999, £164 billion will be spent, with the UK likely to receive about £3 billion. In addition, loans are made by the European Investment Bank (EIB) and the European Coal and Steel Community to areas most seriously affected by agricultural backwardness or industrial decline.

Five objectives have been identified for these funds and govern the eligibility of areas for assistance. Objective 1 is to promote the economic development of regions where per capita GDP is 75 per cent or less of the EU average; objective 2 is to support industrial conversion in regions and local areas seriously affected by industrial decline; objective 3 is to combat long-term unemployment, including pre-vocational and vocational training; objective 4 is to facilitate the occupational integration of young people; and objective 5 is to reform agriculture and forestry, and promote the development of rural areas. For the period 1994–9, Northern Ireland, the Highlands and Islands of Scotland, and Merseyside have been designated objective 1 regions in the UK. Thirty-one areas of Britain, including six London boroughs, will receive objective 2 funding.

Although the Regional Development Fund is restricted to 'objective' or assisted areas, since 1988 the Social Fund has been allocated on a wider basis. Special programmes under 'Community Initiatives' have included RECHAR (for coalmining areas), ENVIREG (environment), INTERREG (cross-border co-operation), KONVER (defence industry conversion and diversification), LEADER (rural development) and the NOW, HORIZON and EUROFORM trilogy (women, disabled or minority groups and vocational qualifications).

European grants to UK local authorities eligible for support from the Regional Development Fund and the Social Fund have provided millions of pounds of public investment. The Social Fund supports many training and retraining schemes, although mainly those within central government programmes. The ERDF has helped to relieve the costs of many infrastructure projects. This fund includes a variety of specific programmes, such as RENAVAL covering shipbuilding areas and PERIFRA for areas affected by cuts in defence spending. The bulk of ERDF allocations has gone to local authority projects. Since 1 April 1993, other public bodies such as UDCs and TECs have also been able to apply for EC structural funds, either in competition or partnership with local authorities.

Grants from the Regional Development Fund are expected to have a genuine additional economic impact in the regions concerned. But until recently the British government used them to reduce its own public expenditure. The grants provide no more than 50 per cent of the cost of a capital project, leaving local authorities to raise the matching funding. Until 1992 spending funded by ERDF grants was counted as falling within a local authority's overall borrowing limit so that no additional spending was allowed. Following pressure from the European Commission, partial additionality was introduced but local authorities still have insufficient funds themselves to take full advantage of ERDF grants. The extent of additionality also varies across government departments.

The amount of European funding available for areas in Britain is likely to decrease within the next few years due to a concentration of resources on economic development in Southern Europe, Ireland and Eastern and Central European countries. However, the British government still hopes to receive more than £3 billion from the structural funds between 1994 and 1999, and is seeking EU aid for the first time for areas of London and the south-east of England affected by industrial decline (*The Guardian*, 12 October 1993). There may also be increasing opportunities for local authorities with European partners to participate in EU schemes to promote small and medium-sized firms, to foster cross-border co-operation in areas ranging from traffic management to tourism development, and to support collaborative research and development (Bongers, 1992). Overall, European funding is likely to diminish in importance compared with the impact of EU legislation on local authority services, making it essential that politicians and chief officers are informed about European matters, rather than these matters being largely confined to staff seeking grants.

The Maastricht Treaty establishes a Committee of the Regions to provide representation for local and regional bodies, with advisory status in relation to the Council of Ministers and the European Commission. The UK is allocated places for twenty-four members on this committee. In the absence of a regional tier of government, the British government adopted the position that UK members of the committee should be appointed by central government and would include local authorities as only one of several 'interest groups'. In March 1993 the government was defeated in the House of Commons on a vote about this issue during the passage through parliament of the European Communities (Amendment) Bill to ratify the Maastricht Treaty. A Labour amendment proposing that all places on the Committee of the Regions be taken by elected local councillors was successful.

Although the British government is not a signatory to the Social Chapter of the Maastricht Treaty, which was adopted by all other member states, many British local authorities have responded to the concept of a 'Social Europe'. As already noted, local authorities have sought and won funding from the European Social Fund, RECHAR (the programme for coalfield communities), the EU Poverty Programmes and the European Development Funds. These funds include objectives for 'social cohesion', 'community solidarity' and tackling 'social exclusion'. They are valuable additional resources for economic and social development in depressed areas and for projects aimed at specific groups such as disabled people and elderly people. However, local authorities have criticised the lack in EU legislation of references to equality of treatment regardless of race, colour or religion, and there is concern that the citizenship rights of minority ethnic groups are not being adequately protected (Bongers, 1992).

Overall, EU developments are having a positive impact for British local government. In contrast to British central government policy, the European Commission emphasises partnership with sub-national levels of government in both policy-making and EU programmes.

CONCLUSIONS

Whilst today the direct state ownership of industry has little support, governments are still expected to intervene in the economy. Legislation and regulations are necessary to protect health and safety and the environment, and training and education are expected to be relevant to the needs of the economy. Economic problems, such as major plant closures, are expected to bring a response from local government and other public services.

Local economic policy is about overall packages of measures, and there is concern that single-purpose agencies such as TECs and UDCs cannot deliver the packages of economic, education, environmental and social measures necessary for economic initiatives to succeed for those people most affected by unemployment. Local authorities therefore have an integrative role in their local economies.

Recent years have seen a trend towards the public and private sectors working more closely together. In Birmingham, for example, the City Council, the TEC and the Chamber of Commerce and Industry have agreed to co-operate within a common economic development strategy (Coopers and Lybrand/Business in the Community, 1992). These partners are integrating their existing enterprise, training and business development programmes. In Nottinghamshire, the 'Common Agenda Initiative' has brought together public and private sector bodies providing business support services. Working groups have been set up to examine inward investment, education and training, and the provision of business support. However, initiatives of this kind are hindered by the number of different bodies involved, often with overlapping roles.

Local economic development is most effective within a strong corporate framework. This is because there are not only important links to be made between key actors in the local economy but also between economic policy and policies for education and training, housing investment, community development and the environment. However, local strategies have limited scope to change the direction of industrial restructuring. The prosperity of peripheral areas of the European Union in particular requires an interventionist industrial policy with social and spatial objectives. The establishment of regional authorities with economic planning and investment powers is increasingly coming to be seen as a key to achieving some control over the direction of economic change, an issue which is returned to in Chapter 11.

9

SUSTAINABLE URBAN POLICY

It is often argued that economic growth is fundamental to both the standard of living and the country's ability to support high standards of public sector provision. However, it is now clear that economic growth has been on an unsustainable course globally and nationally, and that damage is occurring to the Earth's environment and ecosystems. There is a scientific consensus that too little is being done to tackle problems such as global warming, ozone depletion, acid deposition, toxic pollution, species extinction, deforestation, land degradation, water depletion and the poor quality of urban life in many areas. Economic growth must now go together with environmental improvement (Barker, 1993).

A key goal for urban policy is therefore sustainable development and the prevention of damage to the life support functions provided by the environment. Sustainable development is defined by the Brundtland Report of the World Commission on Environment and Development as, 'development that meets the needs of the present without compromising the ability of future generations to meet their own needs' (Jacobs, 1990: 2). At present, urban society is unsustainable because it is both depleting and degrading the quality of environmental resources for future generations. Cities are major generators of gases such as carbon dioxide which cause global warming and of nitrogen oxides and sulphur which contribute to acid rain. The greenhouse effect and ozone depletion are consequences of processes of urbanisation and industrialisation which use up raw materials and energy, and produce damaging waste products. Unsustainable economic growth is endangering the life support functions provided by the environment as a result of the extraction of raw materials and the pollution of air, water and soil.

Threats to the environment occur at all scales, from those which affect the whole planet to local issues (see Table 9.1). The most central issues are those which affect the future viability of life on the planet, but local problems such as litter and traffic noise should not be ignored. They reflect problems with people's attitudes to the environment.

This chapter begins by discussing the scale of change that is necessary and the type of measures at national and local level that can reduce damage to the

Table 9.1 Threats to the environment

Planetary issues	*Local issues*
Generation of carbon dioxide, mainly from burning, causing global warming and potential rise in sea-level.	Changes in the landscape caused by modern farming practices, deforestation or unsuitable afforestation.
Release of CFCs into the atmosphere, causing depletion of the ozone layer.	Damage to ecology caused by pesticides, draining of wetlands, peat extraction, removal of hedgerows and wildflower meadows, over-mowing of verges, park and grassland; lack of planning for wildlife corridors; lack of natural areas for wildlife.
Destruction of habitats, especially tropical rainforests, contributing to global warming and climatic change.	Problems of waste disposal, particularly methane gas, disposal of toxic waste and sewage sludge, and smells from agricultural processes.
Extinction of species.	Manufacture, use or storage of hazardous substances, including asbestos.
Consumption of non-renewable resources and of renewable resources at a rate faster than replacement.	Urban sprawl and the loss of open space.
Land erosion.	The impact of development on urban and rural landscapes.
Human population growth.	Damage to or destruction of trees and woodlands.
National/international issues	Erosion of the countryside caused by over-intensive leisure use.
Radiation from nuclear accidents and waste, and from natural radon gas.	Growth in road traffic generated by new development.
Acid rain, caused by power-station and vehicle exhaust emissions.	Polluting vehicle emissions.
Pollutants in the air and sea.	Noise from traffic, airports, industry, construction, late-night parties and barking dogs.
Water course pollution caused by discharges from sewage works, industry and farming; leaching from waste disposal sites, chemicals in consumer products.	Litter, fly-tipping and graffiti.
Drinking water pollution – mainly aluminium, lead and nitrates.	Dog faeces in public areas.
Pesticides in food.	
Bacteria in food.	

Source: Brooke, 1990.

environment. This is followed by a description of environmental audits and management systems, essential policy instruments for environmental practice. Finally, the links between environmental policy and social and economic policies are discussed.

SUSTAINABLE DEVELOPMENT AND NATIONAL POLICY

Growing concern about environmental problems in recent years has led to an increased awareness that economic growth has to be sustainable. To achieve this end, the consumption of renewable resources should not exceed their regeneration rate, and the use of non-renewable resources such as fossil fuels should not be faster than renewable substitutes can be found for them. Sustainability also requires that the capacity of the environment to absorb wastes is maintained, preventing the build-up of pollutants. Fundamentally, particular features such as ozone levels, atmospheric composition and vegetation cover must be preserved.

In a world still with millions of impoverished people, calling for an end to economic growth to avoid further environmental damage is hardly tenable. However, there is an urgent need to stop growth which exceeds the capacity of the life support functions provided by the environment. Development has to be on a sustainable basis if the long-term health and integrity of the environment is not to be irreversibly damaged. It is a cause of major concern that no economy is yet based on sustainable development. This means that economies are presently borrowing massively from the future. Future generations will have to pay for the resource and environmental degradation accumulating as a result of present industrial and urban processes.

To avert the risk of catastrophic global warming, sharp reductions in carbon dioxide emissions are necessary. The British government has a target of bringing the UK's emissions of carbon dioxide back to 1990 levels by the year 2000, a commitment which arose from the Earth Summit of national governments at Rio de Janeiro in 1991. To achieve this, ministers aim to eliminate a projected increase of 10 million tonnes of carbon a year. The following measures are being used:

1 Increasing duty on petrol, which will save an estimated 2.5 million tonnes.
2 Reducing energy use by central and local government, which will save an estimated 1 million tonnes.
3 Energy efficiency improvements in business, which will save an estimated 2.5 million tonnes.
4 Reducing domestic energy consumption, which will save an estimated 4 million tonnes.

The government expects the extension of Value Added Tax to domestic fuel from April 1994 to reduce energy consumption and has combined this measure with the creation of an Energy Saving Trust, announced in January 1994. The purpose of this trust is to encourage people to buy energy-efficient appliances and insulate their homes. It is proposed that its schemes should be financed by levies on electricity and gas.

Critics argue that more deep-rooted change is necessary (Pearce, 1994). This means thinking on a large scale, such as major investments in district heating from combined heat and power-stations in urban areas, substantial investment in public transport for travel, and land-use policies which enable people to use cars less. In the longer-term, energy must be generated from renewable sources, requiring the construction of tidal barrages and offshore wind generators.

As well as debate about the scale of change that is necessary, there are different opinions about methods. The main issue is whether to use 'market based' methods which rely on pricing and individual decisions or whether to use regulations and direct investment by the public sector. Market-based methods use taxation, subsidies and financial frameworks to make environmentally damaging activities relatively more expensive and environmentally benign activities relatively cheaper. This is sometimes referred to as 'the polluter pays' principle. For example, it is estimated that pollution from road traffic costs about £25 billion a year; this principle would require road users to pay for these costs through taxes on roads and petrol (*The Guardian*, 26 January 1994).

Using the price incentive is gradually becoming more common in the UK, having begun with measures such as taxing unleaded petrol at a lower rate than leaded petrol. However, there is a strong argument that environmentally unsustainable processes should be subject to regulation rather than price disincentives. Regulation works quickly whereas price incentives may take time to have an effect; regulation can effectively ban an environmentally damaging activity or product, whereas pricing only discourages it; and when the product or activity has a low elasticity of demand (demand falls only slowly with price increases), high taxes will be necessary to create a price disincentive (Jacobs, 1990). In addition, the polluter pays principle is not as fair as it might seem:

> the wealthy can often afford to ignore financial incentives whilst the poor can often not afford to respond to them. This raises a more general point. Costs are in general passed on to the consumer. Poorer people may buy less in absolute terms of any particular product but this may still represent a higher proportion of their income. The redistributive effects of making the polluter pay need to be recognised and poorer people either compensated or preferably enabled to avoid the behaviour which is being penalised.
>
> (Association of County Councils *et al.*, 1993: 15)

Pricing, however, does have its place in environmental policy. In particular, it can protect environmental assets by rationing access. Recent concern about visitor pressure in attractive areas of countryside has seen the introduction of pricing. This includes charges for entry to sites and car parks, and tourist 'bed taxes'. These are often used in combination with physical controls to restrict access, such as reducing car park capacities or closing roads. The advantage of charging is that it generates an income for investment in caring for the environment (Bishop, 1991).

Jacobs (1990) argues that public sector investment is essential to realise environmental objectives. Neither pricing nor regulation on their own is sufficient because whole new infrastructures are required to support less environmentally damaging activities. York is a small-scale example of this approach. The city's historic urban fabric has been under threat from growing volumes of traffic. The City Council responded by prioritising infrastructure schemes for pedestrians, cyclists and disabled people, followed by public transport users and short-stay parking for shoppers (Georghiou, 1991). Lowest priority is given to car users. On-site parking in the city centre is restrained and developers have to make up the shortfall with commuted payments towards providing park-and-ride facilities on the edge of the city.

Although the extension of 'polluter pays' price disincentives is the present government's favoured approach to environmental improvement, current environmental protection measures in the UK are by and large regulatory. Such regulations include a ban on smoky fuels in built-up areas, the requirement that all new cars must be capable of using unleaded fuel and the European directive that all new cars must have catalytic converters.

In recent years, in fact, European legislation has been a driving force behind environmental policy in the UK as it attempts to harmonise environmental standards across countries (Bongers, 1992). Well over a hundred environmental directives are already in force, including a series of requirements on water quality, air pollution, substances which deplete the ozone layer, greenhouse gas emissions and waste management (the Commission has taken legal proceedings against the UK government for failure to meet water quality requirements). Major industrial, infrastructure and agricultural projects must have an environmental impact assessment completed before they are approved by planning authorities.

The European Commission's fourth environmental action programme ran from 1987 to 1992 and included a wide range of activity to combat pollution and protect the environment. Its fifth environmental action programme runs from 1993 to 2000. This has a series of targets for industry, energy, transport, agriculture and tourism which are to be addressed with legislation, economic and fiscal incentives and disincentives, public awareness-raising programmes and funding policies. Local and regional authorities are identified as having a particularly important role, with 40 per cent of the programme to be directly implemented by local government.

Following legislation made in May 1990, the European Union will have a European Environment Agency, although implementation has been delayed by controversies about its powers and location. This will provide comparative environmental data and technical and scientific support to member states.

In July 1992 the British government announced plans for an Environment Agency for England and Wales, unifying pollution control functions in one agency by April 1995. This will be the primary UK partner of the European Agency. It will bring together the functions of the National Rivers Authority, Her Majesty's Inspectorate of Pollution and the waste regulation functions of local authorities. The aim is to achieve a fully integrated approach to pollution control. Responsibility for domestic air pollution control under the Environmental Protection and Clean Air Acts will remain with local government. The functions of the Drinking Water Inspectorate, and of local authorities in respect of protecting the health and safety of water consumers, are not being transferred to the new agency.

SUSTAINABLE DEVELOPMENT AND URBAN POLICY

The most urgent need for policy change is at national level. The UK needs a national energy policy based on long-term environmental considerations; long planning horizons and low discount rates for energy conservation and renewable energy investments; tougher building regulations; greater investment in public transport; and the creation of markets for environmental products through regulation, taxation, subsidy and government expenditure. A package of measures was announced by the British prime minister in January 1994, following commitments made at the Rio Earth Summit, but these were widely criticised as inadequate (Pearce, 1994).

The local authority associations have argued that the UK should have a National Sustainable Development Plan (Association of County Councils *et al.*, 1993). Similar to the Netherlands' National Environmental Policy Plan (see p. 233), this would describe sustainable development objectives, the basic principles of action, the required policy directions, targets, and the key policy commitments of the government. The associations argue for the 'round table' approach used in Canada and the Netherlands, bringing together government departments, local government, businesses, non-governmental organisations and academic institutions to work on the plan.

Despite the weaknesses of current national policy, there is an important environmental agenda for public bodies at local level. Cities can be environmentally unfriendly places. Particle and lead concentrations in the atmosphere can cause local pollution, urban traffic is noisy, dangerous and demands infrastructure which disrupts local neighbourhoods, and local problems such as dereliction, litter and high crime rates affect the quality of urban life. These problems have contributed to the movement of firms and people out of inner cities to green field sites, compounding inner city decay and contributing to urban sprawl.

Local authorities can use their direct powers and responsibilities both to enforce and to encourage good environmental practice. Environmental issues cut across virtually every service boundary. Local authorities also have a major role in influencing public opinion, educating and lobbying. Of the threats to the environment listed in Table 9.1, local authorities are usually the lead agencies in addressing local issues, but are increasingly responding at a local level to national, international and global issues.

The rise of environmentalism during the 1980s led to a number of key policy statements and policy guides relating to local government. Friends of the Earth published its *Environmental Charter for Local Government* in 1989. In 1990, the same year that the government published the White Paper *This Common Inheritance* on environmental policy, the local authority associations published a guide to *Environmental Practice in Local Government* (Association of County Councils *et al.*, 1990). At the 1991 Rio Earth Summit a number of proposals were agreed under 'Agenda 21' relating directly to local government functions. These apply to waste management, energy conservation, the integration of land-use and transport planning, protection of natural habitats and coastal planning. The conference proposed that by 1996 most local authorities in each country should have carried out a consultative process with local residents to develop a 'Local Agenda 21', together with targets and timetables (Local Government Management Board, n.d.). There is a particular emphasis on involving women and young people in the planning and implementation of Agenda 21 projects. The measures include:

1 Establishing a community consultation process that brings together representatives of community organisations, industry, business, professional organisations, trade unions, voluntary organisations, educational and cultural institutions, the media and government to create 'partnerships for sustainable development'.
2 Regular environmental audits involving all parts of the community and development of data banks on local environmental conditions.
3 Participation by local councils in regional and international networks of local authorities to increase information sharing and technical assistance and to press national governments to support and fund their environment and development goals.

The Friends of the Earth's *Environmental Charter for Local Government* was sent to all local authorities and has been influential in shaping local government policies and actions. It contains 193 recommendations to promote the conservation and sustainable use of natural resources, and to minimise environmental pollution. Many local authorities have adopted the Friends of the Earth 'Declaration of Commitment' which states:

> This authority will seek to promote the conservation and sustainable use of natural resources and to minimise environmental pollution in all of its own activities, and through its influence over others. This

authority will review all of its policies, programmes and services and undertakes to act wherever necessary to meet the standards set out in this charter.

(Friends of the Earth, 1989)

The charter sets out six policy commitments:

1 To develop an energy policy based on energy conservation and clean technology, and to establish cross-departmental energy management units in local government.
2 To develop a recycling policy that includes provision of public collection or deposit facilities, a recycling officer, an in-house recycling scheme, a commitment to using recycled materials, and either paying or lobbying for rebates paid when recyclable material is taken out of the waste stream.
3 To develop a strategy for both monitoring and minimising pollution in the local environment, including pollution caused by local government's own activities. This should make use of all available measures, including publicity and enforcement.
4 To develop transport and planning policies to minimise the use of cars and encourage public transport, cycling and walking. Fuel efficiency, the use of unleaded fuel and the fitting of catalytic converters to minimise pollution from vehicle emissions should also be encouraged.
5 To develop an environmental protection and enhancement strategy, including measures to protect and enhance public open spaces, wildlife habitats and streets. Full consideration should be given to the particular needs of disadvantaged groups. Environmentally sustainable methods of land management (which do not threaten the enjoyment of wildlife) should be adopted for parks, open spaces, and verges, and organic and sustainable methods in agriculture and countryside management should be promoted.
6 To develop a health strategy which recognises the links between the environment and public health and includes implementation of a health and safety policy which has full regard to environmental hazards.

The Friends of the Earth charter contains a large number of recommendations, covering every aspect of how a local authority operates and functions. This comprehensive approach is also promoted by the Environment City scheme.

In 1991, Leicester was designated Britain's first Environment City by the Royal Society for Nature Conservation and the Civic Trust. Leicester Environment City itself is an independent charitable trust with the aim of moving the city towards sustainability by working through partnerships with the public, private and voluntary sectors. The programme has been developed around eight broad themes: energy, transport, waste and pollution, food and agriculture, economy and work, the built environment, the natural environment, and the social environment. Specialist working groups tackle each of the themes, with representation from public, private and voluntary organisations. Task forces focus on specific issues.

A key aspect of Environment City is a strategic vision for Leicester called Agenda 2020. The strategic objectives of Agenda 2020 include 'green' and safe urban development; repopulating the city centre; energy conservation and the use of clean and renewable energy technologies; reducing the need to travel and prioritising alternatives to the car; urban nature conservation; encouraging sustainable and healthy diets and systems of food production; work within schools and the community to raise environmental awareness; and sustainable waste disposal and pollution control policies (see Appendix 1). Many projects are already under way. These include tree planting, a ban on the use of peat by the City Council, guidance for environmentally friendly housing design and construction, combined heat and power schemes, traffic calming and traffic free zones, and environmental education packs for schools (see Appendix 2). Leicester City Council has a key role in Environment City and in 1992 established a new chief officer post of Director of Environment and Development.

Other Environment Cities in Britain are Middlesbrough, Leeds and Peterborough. These cities are setting new standards in key areas of environmental policy. Middlesbrough, for example, has set up an air pollution monitoring network which is probably the most extensive in Europe (*Environment Forum*, No. 4, July 1993). Regular pollution maps are published and feedback from residents is sought to help tackle pollution problems. Leeds has initiated one of the largest household recycling projects in Europe, with nearly 90,000 homes involved in a kerbside recycling scheme. Two 25-storey point blocks have been transformed into 'tower blocks of the future' with re-cladding, re-roofing, double-glazing and a combined heat and power plant, saving energy, cutting tenants' heating bills and reducing carbon dioxide emissions by two-thirds. Peterborough has introduced a cycling strategy to reduce car use, and has about 10–12 per cent of commuters travelling to work by bike. Environment Cities hold their title for four years and then have to bid again in competition against other cities to retain their designation.

Overall, there are six main areas where there can be important local authority contributions to sustainability: transport policy, energy, waste management, utilities, green spaces and green purchasing.

Transport policy

One of the most pressing issues for environmental policy is to achieve sustainable mobility. In 1993, the European Commission published a White Paper, *The Future Development of the Common Transport Policy: a global framework for sustainable mobility* (COM 92 494 final), which identified energy consumption, pollution, congestion, effects on land use and land take, and the carriage of dangerous goods as the main environmental problems facing European transport policy.

In 1990, road transport accounted for 93 per cent of all passenger kilometres in the UK and 63 per cent of freight tonne kilometres (Association of County Councils *et al.*, 1992). It generates 19 per cent of total emissions of carbon dioxide, one of the main greenhouse gases. In 1989, the Department of Transport estimated that traffic on UK roads would increase by between 83 per cent and 142 per cent by 2025. Despite improvements in fuel efficiency, this would result in carbon dioxide emissions increasing by between 60 per cent and double.

Scientists and environmentalists have argued for some time that unacceptable damage to the environment is being caused by the growth in road traffic. In January 1994, the government acknowledged that this growth could not continue unchecked when the prime minister announced measures to follow commitments made at the Rio Earth Summit. This new thinking represents a significant victory for the Department of the Environment over the Department of Transport, which has promoted road building to cater for the projected growth in traffic (Hamer, 1994). The government has now committed itself to expanding traffic calming and giving higher priority for government funding to facilities for public transport and for walking and cycling, as well as higher 'environmental' taxes on energy and roads. It has also acknowledged for the first time that further action to cut carbon dioxide emissions will be needed after the turn of the century when they are predicted to rise again.

Car transport is not only an environmentally damaging way of moving people about cities, it is very inefficient. Innovations such as catalytic converters, quieter engines and computerised traffic lights have sought to reduce pollution and noise from motor cars but do little to limit traffic growth. Traffic restraint is unlikely to be achieved without radical policy changes.

Road pricing is attracting a growing amount of support as a way of reducing congestion, using the revenue raised to maintain roads and support improved public transport. Cambridge introduced an experimental road pricing system in 1993. It works by drivers who wish to use a designated metered zone having to install pre-paid smart-cards in their vehicles. These activate radio beacons in congested conditions. Cars entering the metered zone are only charged if conditions are so congested that they stop more than four times within any 500 metres or if they take more than three minutes to travel this distance. Revenue raised from the scheme is used to improve public transport. Within the city, different modes of transport are separated as far as possible, with priority being given to pedestrians. There are different routes for pedestrians and cyclists, bus lanes, and park-and-ride schemes for cars.

Road pricing may not be the answer, however. Research commissioned by Humberside County Council concluded that road pricing in Hull would lead to a significant deterioration in air quality as drivers diverted around the defined pricing areas, also drawing economic development away from the centre of the city (Association of County Councils *et al.*, 1992).

A key goal of sustainable urban policy must be to reduce dependence on cars. Although fitting catalytic converters will reduce the environmental impact of each individual car and raise its *environmental efficiency*, it will not maintain *environmental capacity* if car use continues to grow. Emissions from rising car use are projected to increase over the next ten years at a rate which will exceed the rise in environmental efficiency resulting from catalytic converters (Jacobs and Stott, 1992). But it is unlikely that public transport will ever substitute for many trips now made by car. The aim of public transport policy must be to target improvements where they can be an efficient alternative to cars. Other measures will be needed to enable people to use cars less.

Newman and Kenworthy (1989) distinguish between 'automobile cities', 'public transport cities' and 'walking cities'. They found that the less dependence a city population has on automobiles, the higher are its population densities (which support local facilities), the lower is the provision of central business district parking, the higher is city-wide public transport provision, and the lower is road provision per capita. Urban planning will need to generalise the characteristics of 'public transport' and 'walking' cities.

Roberts (1991) argues that properly funded and high-quality public transport, walking and cycling are the foundations of development which has least impact on the environment. A major concern in this respect is the amount of car-based shopping centres which today generate a considerable amount of traffic. Roberts argues instead for a hierarchy of shopping areas related to where people live and work, with everyday shopping made accessible on foot, no more out-of-town shopping centres, and encouragement for shopping from TV data systems at home, with deliveries by retailers.

Evidence from German towns suggests that reducing parking spaces has no effect on retail turnover and can actually increase it when combined with improvements in public transport. In 1989, the Netherlands government's Ministry of Housing, Physical Planning and Environment published a National Environmental Policy Plan, *To Choose or to Lose*. This is a comprehensive action plan for sustainable development. In aiming to reduce harmful emissions from cars and trucks, the plan requires that, 'the locations where people live, work, shop and spend their leisure time will be coordinated in such a way that the need to travel is minimal' (Government of the Netherlands, 1989: 195).

Sustainable land-use policy involves locating urban development in ways that encourage shorter and less polluting journeys. Mixing different land

Table 9.2 Key results from household surveys about travel patterns in Milton Keynes and Almere

	Milton Keynes (%)	*Almere (%)*
Proportion of trips made by car	69	43
Proportion of shopping trips made by car	65	58
Proportion of households without a car	10	21
Proportion of households with two or more cars	39	14
Proportion stating car is essential to their lifestyle	71	52
Proportion of households who never let their children out unsupervised (an indicator of risk of traffic accidents)	52	16

Source: Rawcliffe and Roberts, 1991.

uses and activities together in an urban area is better than strict, segregated land-use zoning. It allows people to work, live and take part in leisure activities without the need to travel long distances. Areas of relatively high density, particularly associated with high capacity public transport systems, are more efficient than developments such as low-density suburban housing with separate employment areas and poor public transport. High-density, mixed development is also conducive to the development of combined heat and power schemes.

Rawcliffe and Roberts (1991) compared two cities with different land-use and transport planning, but which were otherwise very similar, in order to investigate how this affected travel patterns. These were Milton Keynes in Britain and Almere in the Netherlands. The different physical planning of the two towns is summed up as follows:

> [Milton Keynes] was designed on a grid-iron framework to accommodate the car, with cycleways slotted in later. Almere has a more balanced transport system – segregated busways, segregated cycleways and walkways – which is not geared towards local car movement; access roads are narrow, with speed humps every 50 metres or so.
>
> (Rawcliffe and Roberts, 1991: 310)

Table 9.2 presents some interesting findings from household surveys in the two towns. The different land-use patterns appear to have a marked effect on characteristics of travel, with the reduced distances to urban facilities and the greater significance of walking and cycling facilities in Almere resulting in more sustainable travel patterns.

Local councils can pursue environmental traffic policies in a number of ways, from the adoption of physical measures such as ramps and chicanes to promote safety and improve the environment under the Traffic Calming Act 1992, to comprehensive strategies for linking land-use and transport planning to environmental objectives. Improving the quality and availability of public transport is absolutely central to this. Leicestershire County Council, for example, has a 'transport choice strategy', the main elements of which are:

1 A complete county rail network with County Council investment and profit-sharing with British Rail.
2 County Council investment in bus priorities, bus terminals and bus information systems. Partnership with bus operators to use additional revenue from these investments to achieve higher-quality vehicles and higher standards of operation. Marketing of bus services as a quality alternative to car use.
3 New land-use policies directing new development only to locations where there is a choice of high-quality bus or rail transport.
4 Investment in extensive traffic calming schemes, with complementary routes for pedestrians and cyclists.

In summary, a sustainable transport policy for cities has three key components:

1 Reducing the overall length of trips.
2 Encouraging public transport use when travel is necessary.
3 Adopting a policy framework that gives walking and cycling precedence (within their limits) over other forms of travel.

Energy

Urban policy can complement national policies for sustainability in energy supply and energy use. In land use and construction, consideration of passive solar energy and micro-climatic design can be encouraged by planning authorities. Careful building design can increase the benefits of solar gain and avoid any negative effects of micro-climate. Energy consumption is reduced by designing buildings to face south, with larger windows towards the sun and small windows to the rear. Buildings can also incorporate heat-retaining features. Computer systems can be installed in buildings to monitor energy use and control more accurately heating, ventilation and air conditioning. Local authorities can incorporate these features in their own new or refurbished buildings, although they cannot use their planning powers to require energy measures in private developments, and there is no statutory requirement for them to base plans on environmental targets.

Local authorities, nevertheless, use their planning and economic development functions to encourage energy initiatives in the private sector. The following are the main examples:

1 Combined heat and power stations. These produce electricity and useful heat. There are few large-scale city-wide combined heat and power plants in the UK, but small-scale CHP units (such as for swimming pools

and housing complexes) are an effective way of reducing carbon dioxide emissions. They increase energy efficiency from around 30 per cent in conventional power stations to up to 80 per cent.

2 The use of passive solar energy designs in new and refurbished buildings.
3 Use of renewables such as photovoltaic roof and wall panels, landfill gas, bio-gas from farm waste, energy crops, and wind and hydro power.
4 Creation of Energy Conservation or Action Areas which target action to bring housing and business premises up to a higher standard of energy efficiency.

Energy efficiency is an area where much more central government action is required. There is a pressing need for a national energy policy based on long-term environmental considerations. This should also include creating a financial framework which permits longer payback periods on energy conservation and renewable energy investments than existing accounting rules in the public sector allow.

Waste management

Local authorities have made important contributions to the sustainability of waste management. Central government policy is that 50 per cent of *recyclable* household waste should be recycled by the end of the 1990s (Department of the Environment, 1990). Recyclable waste represents about 25 per cent of total household waste; currently only about 2.5 per cent of all waste is recycled.

Local councils have responsibility for collecting and disposing of waste. Section 49 of the Environmental Protection Act 1990 places a duty on local councils to produce recycling plans. These are strategies for making less use of landfill to dispose of waste, and greater use of recycling facilities for glass, cans, paper, plastic, metal, oils, compostables and textiles, and of incinerating waste to produce power.

Survey evidence suggests that more collection points for recycling waste are necessary if the current figure of about 40 per cent of people who regularly take bottles, cans or newspapers to these points is to be increased (*Hansard*, 1992d). Lack of storage space in the home is one factor behind the present relatively low level of use. Some councils now have kerbside collections which require residents to carry out some sorting of their waste before collection. These reduce the number of cars travelling to recycling points and increase participation in recycling to 70–80 per cent, but the schemes are costly to operate. Deposits on bottles have been found to have a marked effect in encouraging people to separate items for re-use.

Newcastle City Council contracts with the Byker Reclamation Plant which produces refuse-derived fuel pellets from the paper and plastic elements of household waste. Ferrous metal is also recovered from this process. The concept for the Byker scheme originates from a central government initiative in the mid-1970s, 'War on Waste'. A market for the pellets was created by the development of an adjoining combined heat and power plant which provides heat for the nearby Byker housing development. With the Electricity Act 1989 and the privatisation of electricity, a market was also created for electricity generated from burning the fuel pellets.

However, recycling processes use energy and can produce hazardous residues. It is not a substitute for conserving resources and using less energy. There is a need for measures which minimise waste in the first place, such as reducing packaging and re-using containers. But, as in other areas, financial restrictions on local councils severely limit how much they can invest in measures of this kind. There is also wide agreement that central government action is necessary to ensure a market for recycled products, such as taxing non-reusable containers, making the purchase of recycled goods obligatory for public bodies and specific sectors of industry, introducing tax relief for the capital costs of recycling plants, and requiring 'eco-labelling' so that consumers are informed of the 'green quality' of products.

Utilities

In urban areas, the most fundamental aspect of environmental quality is the life support services provided by infrastructure networks for water, sewerage and energy services. In 1914, local authorities were responsible for 80 per cent of water sales, 68 per cent of electricity sales and 37 per cent of gas sales. They undertook all sewerage collection and treatment, provided tram networks and telephone systems (Marvin, 1992b). Local government ownership and control of these systems largely ended with the nationalisation of the major utility networks in the 1940s. During the 1980s these networks were privatised.

Today, millions of households have major difficulties paying for sufficient energy to heat and light their homes. During 1990/1, almost 1 million summonses were issued for non-payment of water charges. Sharply rising water charges to pay for new environmental standards are coinciding with the introduction of water metering, raising concerns that those on low incomes with needs for high water usage due to illness or family size will be unable to meet their water bills.

These services are a neglected area of urban policy, despite their importance to life support and sustainable development (Marvin, 1992b). They are also necessary for a competitive local economy and are key investment targets for measures to conserve resources. There are marked inequalities in access to them, with poor access, underinvestment and the costs of inefficient systems tending to concentrate in deprived areas where they are needed most for economic competitiveness. A similar situation exists regarding low levels of connection to telecommunication networks. Affordability and disconnection from services such as water and electricity are overwhelmingly problems for low-income households. In deprived areas, high densities of disconnections are likely to threaten serious health and environmental problems. Low-income households are often doubly disadvantaged by living in homes that are difficult to heat because of poor insulation or damp, and by needing to keep warm for reasons of disability, old age or young children.

Marvin argues that although local authorities no longer own or control most infrastructure networks, they still have an important role in promoting co-ordination between the networks and reducing inequalities in access to them. For example, ducting for cable and district heating networks can be included in new road projects to attract further investment to priority areas; local authorities can research the extent of fuel poverty and press the utilities and regulators to provide an improved level of services which recognises these needs; they can negotiate cheaper tariffs or energy efficiency schemes from the utilities for council tenants; and they can enter into partnerships such as combined heat and power/district heating schemes or light rail transport networks. There is, Marvin argues, a need for cities to identify their infrastructure problems and to return the provision, performance and quality of infrastructure services to an urban policy framework.

Green spaces

An important local contribution to environmental policy which local authorities can make is promoting and enhancing their green resources. Many urban areas have significant amounts of green space. Newcastle City Council, for example, has a countryside strategy which includes large areas of green belt around the city, and areas of woodland, wetland and parkland within the city. It has five sites of Special Scientific Interest, twenty-four sites of Nature Conservation Importance, twenty-five Sites of Local Conservation Interest, seven Local Nature Reserves, four Ancient Woodlands, five Community Woodlands and eighteen Sites of Potential or Known Archaeological Interest. A corporate strategy has been developed to protect and enhance these resources whilst increasing public use and understanding of the countryside (Newcastle City Council, 1991).

Tourism has an important contribution to make to raising environmental awareness. Many local authorities are now planning for sustainable tourism, a long-term approach to tourism development which seeks to balance the needs of visitors with those of wildlife, landscape and buildings. 'Eco-tourism' is a growth sector, promoting environmentally friendly breaks and holidays.

Green purchasing

The public sector spends billions of pounds a year purchasing materials and goods. It is estimated that local authorities spend about £3 billion per annum on buying goods and services. Purchases are mainly building materials, food, energy, vehicles, road materials, office supplies, uniforms, protective clothing, street furniture, cleaning materials, horticultural products and office furniture. Production of all these goods can involve environmentally damaging processes. Although public sector purchasing is normally subject to competitive tendering and other regulations to ensure value for money and fair competition, the nature and means of production of goods can be specified by public sector purchasers. Many public sector organisations, especially local authorities, are now purchasing goods that minimise damage to the environment, either in their use or in their production and distribution (Holmes, 1993).

ENVIRONMENTAL AUDITING AND MANAGEMENT

Environmental auditing of key environmental features is fundamental to environmental analysis, policy formulation and monitoring. Local councils have pioneered this practice in the UK. However, they are not required by law to conduct audits and probably less than a hundred do so, despite a strong case that all should institute regular auditing as part of a corporate environmental management system (Association of County Councils *et al.*, 1992).

Britain has lagged behind some other countries in adopting this approach, although the 1990 White Paper *This Common Inheritance* established a series of environmental objectives which are monitored through annual progress reports. Canada adopted national state of the environment reporting in 1986 and now has a sophisticated system for presenting information on 'what is happening' (environmental trends and conditions), 'why it is happening', 'what are the impacts', and 'what is being done' (Selman, 1992).

A local environmental audit can be one of three kinds (Taylor, 1992):

1 A *State of the Environment* audit. This analyses environmental conditions in the local authority area.
2 An *Internal Audit* or *Review*. This investigates the environmental impacts of the council's own policies and practices, for example:
   (a) Does the authority have an active energy conservation programme?
   (b) Is lead-free petrol used?
   (c) Do all new petrol vehicles have catalytic converters?
   (d) Is the recycling of materials actively encouraged?
   (e) Are purchases of CFCs and harmful pesticides avoided?

   As well as reviewing internal practices, an internal audit can also assess the authority's overall policies from an environmental perspective. For example:

(a) Is the whole district a smoke control area?
(b) What is the balance between conservation and development?
(c) Is the impact of traffic on residents examined and are traffic calming measures considered?
(d) Is land maintenance carried out with regard to ecological considerations?
(e) Have conservation areas been introduced?
(f) Is there an active reclamation policy for derelict land?
(g) Are areas of natural beauty and Sites of Special Scientific Interest fully protected?

Internal audits can be more fully developed to detail the environmental costs and benefits of policies and practices. From this, priorities for action can be identified on the basis of environmental benefit, severity of the problem, and financial cost (although cost will only be part of an equation which will include other measures of value).

3 A *Management Audit*. This reviews the procedures used by the council to carry out its environmental policies, commitments and responsibilities. Brooke (1990) argues that there must be a corporate policy, with the commitment of politicians and involvement of the chief executive, which ensures that environmental factors are taken into account in all decision-making.

Lancashire County Council was one of the pioneers of environmental auditing in Britain and has produced all three kinds of audit. To carry this out, good research, data processing and data presentation have proved essential.

Lancashire called the first kind of audit its *Green Audit*. This had two main aims. First, to give people and organisations in Lancashire an overall picture of the state of the local environment; second, to use this knowledge to define the actions needed to sustain and improve the state of the environment. Within these aims, there are specific objectives for the audit. For example, the following objectives were identified for research and information:

1 To produce a comprehensive analysis of environmental conditions using specific indicators of environmental quality developed from the most recent publicly accessible sources of data.
2 To establish a baseline against which environmental change can be monitored.
3 To create an environmental database.
4 To identify gaps in existing data which need future action.

For the purposes of the audit, the environment was broken down into more manageable components. These were: structure (geology, soils, etc.);

air; water; waste; noise; energy; land and agriculture; wildlife; landscape and townscape; open space; and transport. Measurable attributes of each of these components were identified and used as indicators. For example, levels of pollutant gases were used as indicators of air quality, and an in-house scenic quality survey provided indicators of landscape quality.

The planning and management of the audit was organised as follows:

1 The scope, objectives and time-scale of the audit were determined by the Lancashire Environment Forum, chaired by the leader of the council and comprising some seventy member organisations in local and central government, industry, higher education and the voluntary sector. Most of the data for the audit were supplied through members of the forum. It was also given the role of commenting on drafts and publicising final documents. A smaller steering group was established to act on decisions taken by the forum.
2 An officers' management group chaired by the council's chief executive was established to manage the audit.
3 An environmental unit of five specialists located in the planning department was established to carry out day-to-day work.
4 A computer development group was formed to develop information technology requirements.
5 It was decided that the council's policy and resources committee, chaired by the leader of the council, should receive reports based on the audit.

The indicators were used to give a composite picture from which to identify issues and make an action programme. The action programme was drawn up by specialist working groups of the forum.

A key requirement for the audit was a system for *processing information*. Five key requirements for such a system were identified:

1 It would have to hold, access and manipulate a lot of different information which had one common feature: it would relate to places on, under or above the surface of Lancashire.
2 It would have to show what exists in particular places, using a wide range of quantitative and qualitative attributes, and to identify places where particular environmental conditions applied.
3 It would have to handle time-series data for monitoring change over time.
4 It would have to support as wide a range as possible of environmental problem-solving and policy-analysis situations.
5 It would have to output information for geographical areas and in ways that could be easily understood.

These requirements led the council to adopt a geographical information system (GIS) solution. An ARC/INFO GIS kit was purchased.

Before data input to the GIS could commence the ground had to be properly prepared. There had to be time to develop skills using ARC/INFO. The required data had to be identified, their existence verified, the data holder contacted, the format of the data established, and confirmation that they could be released had to be obtained. The data had to be structured within the framework of the components of the audit and its indicators. There had to be time to gain familiarity with the nature and scope of the data and establish working relationships and practices. Basic map scales and geographical boundaries had to be determined.

Inputting the data was very time consuming. Data had to be correctly formatted before input could begin, and for some data formatting took longer than inputting. In total, assembly of the GIS database took fourteen months. Once assembled, it became a powerful information resource which could be routinely updated. Data could be presented at a wide range of different geographical scales. Data coverages included sources of air pollution and contaminated land locations, solid geology, river quality, land liable to flood, soils, agricultural land quality, census data, drinking water quality, landscape character, planning policy designations, habitat survey data, ancient woodlands and mineral workings.

The following applications for output from the GIS were established: (a) to develop policy measures; (b) to create public access to environmental information, enabling access to the data from public libraries and schools using CD Rom versions of the database for use with microcomputers; (c) to support reviews of land-use policy, energy policy, and minerals and waste policy.

In April 1993, the Lancashire Environmental Forum published the Lancashire Environmental Action Plan, a local Agenda 21. Updated annually and reviewed every five years, the plan will also be informed by updates every three years of the environmental audit.

Many other local councils are conducting environmental audits. Newcastle City Council (1989) has identified a number of environmental indicators across several areas of activity (see Table 9.3). The indicators were adopted to monitor the success or failure of specific city council policies, to demonstrate where there is a need for further action, to provide information on the state of the environment, and to promote better understanding of the environment. For each indicator, a city standard is identified when available and action related to the indicator is outlined. For example, the target for buying recycled or 'environmentally friendly' products was set at 100 per cent of purchases and the target for energy conservation was set at a 15 per cent reduction in consumption using capital investment and careful management. For some areas of activity, special surveys were used to assess the extent and nature of problems. An infra-red aerial survey of Newcastle was used to obtain a visual image of where heat was escaping from buildings, enabling remedial action to be targeted on heat leaks and poorly insulated roofs.

Table 9.3 Newcastle City Council environmental audit indicators

Area of concern	*Indicators*
Energy conservation	Energy efficiency standard of city buildings. Energy saving measure for housing.
City Council buying policy	Percentage of goods purchased which are recycled or environment friendly.
Public health	Standardised mortality ratio. Birthweight. Rates of permanent sickness or disability. Dog nuisance measures. Food testing and visits to food premises. Complaints of rat infestation and programme of poison baiting in sewers.
Air pollution	Smoke/sulphur dioxide levels. Nitrogen dioxide levels. Atmospheric lead levels. CFC removal. Radiation levels. Noise nuisances.
Water pollution	Lead content. Nitrate levels. Aluminium levels. River quality.
Transport conditions	Highway maintenance standards. Accident levels. Congestion measures.
Waste management	Percentage of waste recycled. Street cleanliness standards. Ground pollution.
Recycling	Number of facilities. Number of recycling initiatives.
The living city	Distribution of flora and fauna. Open space standards. Landscape standards. Reclamation of derelict land and amount remaining.

Source: Newcastle City Council, 1989.

Environmental audits are an important means of providing a sound basis of information about trends in environmental indicators and key environmental pressures. However, there is a need for the various indicators used by different local authorities to be standardised within a national framework for information collection, including both national coverage and updating. The UK local government associations have argued that this should include establishing a set of sustainability/'sustainable lifestyle' indicators which local authorities should collect and report. They have also advocated research into a 'quality of life' index that might also be published (Association of County Councils *et al.*, 1993).

Environmental audits are strategic tools. Major development projects now have to be audited separately for their environmental impact. This is required by the European Community Directive 85/337, which is incorporated into British law by the Town and Country Planning (Assessment of Environmental Effects) Regulations 1988. *Environmental statements* prepared under these regulations include:

Table 9.4 Extract from West Central Route Environmental Statement

Landscape:	
Positive impacts	*Negative impacts*
1 Total of 649 extra heavy standard trees will be planted along the West Central Route corridor.	1 154 mature trees lost along the West Central Route corridor.
2 22,360 new shrubs are to be planted in and adjacent to the road corridor providing a total planted area of 5,590 sq m.	2 The grading of Hunters Moor will interfere with the natural form of the moor.
3 Closure of Ponteland Road and Grandstand Road (parts of) will remove vehicle related stress from approximately 300 mature trees on Nuns Moor.	3 Possible damage to soil structure by livestock could have a long-term detrimental effect on the existing trees at Nuns Moor where fences are removed.
4 The regrading of Hunters Moor will reduce the impact of the existing embankments at the north end of Claremont Road.	

Source: Newcastle City Council, 1992a.

1 A description of the proposals, comprising information about the site, design and scale of the development.
2 The data necessary to identify and assess the main effects which the scheme will have on the environment.
3 Descriptions of the significant effects, direct or indirect, of the development on the environment.
4 Descriptions of possible measures envisaged to avoid, reduce or remedy any significant adverse effect.
5 A summary, in non-technical language, of the above information.

Newcastle City Council prepared an environmental statement for the authority's proposed 'West Central Route'. This road comprises part of the overall transport strategy for the city centre, and is intended to provide a traffic corridor through existing commercial, housing, recreational and green space areas, relieving congestion and improving conditions for pedestrians, cyclists and public transport. The environmental statement includes a description of the proposals and their objectives, followed by a 'balance sheet' of the positive and negative environmental effects of the scheme, including the construction impact and the impacts on vehicular emissions, noise, the visual and built environment, landscape, ecology, recreation and employment. The road would cut through part of the Town Moor and Hunters Moor, green space used predominantly for grazing. Table 9.4 shows how the impact on this landscape

Figure 9.1 BS 7750 model for environmental management in an organisation
Source: Sheldon, 1992.

was evaluated in the environmental statement.

Environmental policy has to have a management process to make things happen. A corporate approach is essential because many policies and objectives rely on individual services working together. The local authority associations recommend management by a cross-service steering group which cuts across the hierarchies of conventional departmental organisation:

> To manage the environment effectively means a break with hierarchical management styles, and the development of teamwork and cross-service initiatives ... Basically this means having a cross-service and cross-committee decision-making structure. A green action steering group or state of the environment working party are good models ... containing a wide range of officers from the top to the bottom of the structure, but united by an interest in and responsibility for the environment. Both officers and members, together and separately, need a suitable forum.
>
> (Association of County Councils *et al.*, 1993: A4)

A British Standard has been developed for environmental management systems. In the same way that the award of BS 5750 to an organisation assures the quality management system which the organisation operates (see Chapter

5), a new British standard, BS 7750, assures the environmental management system of an organisation (Sheldon, 1992). Organisations subscribing to the new European Eco-management and Audit Regulation will be required to introduce such systems. They are designed to bring about a significant impact on environmental performance, both in the public and private sectors. Registration to BS 7750 involves introducing a system for the management control of environmental practices and the assessment of compliance with the organisation's environmental policies (see Figure 9.1). The standard is a management tool, not an environmental policy or standard.

Raemaekers (1993) argues that environmental management systems are the new third phase of 'corporate environmental management' in local government. The first phase saw the introduction of environmental charters and action programmes in the late 1980s. The second phase saw the appearance of environmental audits and state of the environment reports in the early 1990s. Following the third phase of introducing environmental management systems, Raemaekers suggests that local Agenda 21s mark the start of a fourth phase of development.

ENVIRONMENTAL URBAN POLICY

Economic growth is necessary to bring about improvements in welfare, but this can only make sense if growth is sustainable. Environmental improvements are often presented in terms of costs but there is no evidence that countries with strong environmental policies experience lower economic growth. Indeed, investment in producing new energy-efficient appliances, reducing the energy consumption of buildings and industrial processes, generating renewable energy, recycling materials and cleaning up past pollution is a source of growth and employment (Barker, 1993). For the economy to work in this way, however, it is necessary for governments to create a framework of environmental standards, subsidies, grants and taxes on pollution.

At the local level, Jacobs and Stott (1992) discuss the growth in recent years of environmental policies in local government, but comment that environmental economics has had little influence on local authorities' economic development practices. Jacobs and Stott propose reorientating local economic policy towards sustainable development. They argue that local authority economic development units should be replaced by 'environmental economic development units' which would use information packs, grants and research to encourage waste minimisation and resource efficiency, green purchasing, environmental training, environmental market and sector research, and energy efficiency. Local authority services including land-use planning, pollution control, waste regulation and environmental assistance programmes to businesses could be grouped together in a single department which 'could truly be regarded as a local Environmental Protection Agency' (Jacobs and Stott, 1992: 268).

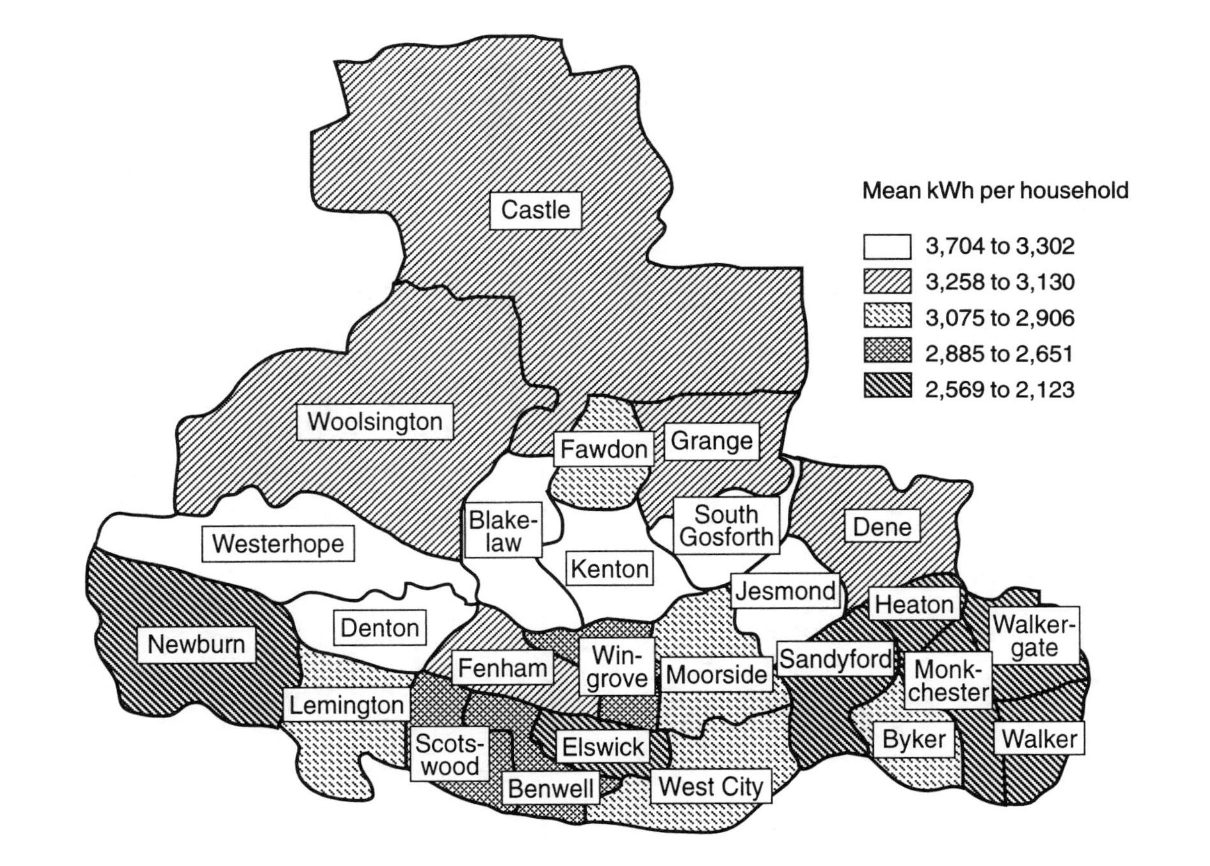

Figure 9.2 Average yearly domestic consumption of electricity, Newcastle upon Tyne, 1989 (from postcode sectors to wards)
Source: Marvin, 1992.

As well as being a major issue for economic policy, possibly demanding the organisational change Jacobs and Scott advocate, sustainable development is a major issue for social policy. Better-off people spend more money on energy-intensive activities such as travel and have more domestic appliances such as dishwashers and tumble dryers than poorer people, who consume less energy (Marvin, 1992a). Figure 9.2 shows how households living in the high unemployment riverside wards of Newcastle consume significantly less electricity than the rest of the city. The scale of resource depletion by higher income groups is largely an unpriced benefit for the better-off; the full environmental costs of their consumption are not reflected in their fuel bills.

However, the effect of introducing measures such as a carbon tax to reduce energy consumption will impose greater costs on lower-income groups, where there is already a substantial problem of fuel poverty (Jacobs, 1990). This is the major criticism of the government's decision in 1993 to extend VAT to domestic fuel, a decision which is unlikely to reduce fuel consumption significantly and was probably motivated chiefly by Treasury pressure to raise revenue to reduce the public sector borrowing requirement. The local authority associations had proposed replacing the present pricing system with a progressive banded tariff that would have a low rate for the first 'subsistence' units and two higher bands for 'luxury' consumption levels (Association of County Council *et al.*, 1993).

The most effective way of protecting low-income groups is to subsidise environmental goods and services from progressive general taxation, even if this is shifted from taxes on employment to taxes on carbon dioxide emissions and damaging activities such as urban traffic. The positive effects of such subsidies are illustrated by many local authority programmes for council housing. Council tenants spend over £2.5 billion a year on their fuel bills. Despite this, many households live in fuel poverty due to low incomes and hard-to-heat housing. Inadequately heated homes also give rise to health problems due to cold, damp and mould growth. It is common practice now for the refurbishment of local authority housing to prioritise energy efficiency, installing insulation and draughtproofing, or carrying out major works such as new heating systems or complete rendering of the shell to improve wall insulation and eliminate damp penetration. These measures also reduce the management and maintenance problems that result from inefficient heating.

Environmental urban policy is essentially a paradigm for urban policy as a whole. Urban policy is about achieving the best distribution of available urban resources to improve welfare. Environmental urban policy adds the dimension of distribution across generations: meeting present needs without compromising the ability of future generations to meet their needs by depleting or degrading the quality of existing resources.

10

HEALTH FOR ALL

Urban policy has its origins in the public health tradition and concerns with controlling environmental causes of disease. These became less important in health policy with the expansion of the National Health Service and its treatment services for diseases and illnesses. However, this emphasis on curative medicine began to change during the 1980s when a revival of public health occurred as a result of new concerns about the effects of the environment, lifestyles and social inequalities on health. An important initiative was the Healthy Cities project promoted by the World Health Organisation in Europe. This defined health in its broadest sense, as illustrated by the following indicators of a healthy city (Hogg, 1991: 16–17):

1 Nuisance indicators, such as levels of noise, odour and cleanliness.
2 The proportion of green space.
3 The percentage of children with criminal records.
4 The employment rate among adults with disabilities.
5 The prevalence of functional illiteracy.
6 Poverty.
7 Poor housing.

Such indicators reflect a very different conception of health compared with the medical view. Instead of seeing health as the treatment of diseases and illnesses, the healthy cities approach has been to work for equal opportunities to enjoy positive health. This has meant addressing the factors which cause damage to health.

Thus, Hogg explains:

> Healthy diet is more expensive than a diet of junk food. Affordable recreation must be available to increase participation in sports and exercise. Smoking is increasingly associated with women and with disadvantage; it seems to provide relief in a stressful life. The circumstances causing the stress need to be addressed. The environment needs to be protected from pollution. Helping the more disadvantaged communities means enabling them to gain control over their lives. The

most effective way to empower people is to ensure that they have a secure and adequate income so that they can participate in the community.

(Hogg, 1991: 78)

This chapter is about public health goals for urban policy. It considers firstly the nature of public health problems in urban society. It contrasts the role of the NHS with the role of local authorities. Four policy areas are examined in more detail: environmental and public protection, housing and planning, health promotion (including leisure services and education), and social services. The chapter then discusses the UK's current long-term health strategy set out in *The Health of the Nation* White Paper. It compares this strategy with the 'health for all' policies that many local authorities and voluntary organisations are pursuing in their local areas. The chapter ends by stressing the importance of health in all public services. As awareness grows that health is about much more than health care services, it is likely that local authorities will assume a much greater role in health strategy, possibly extending to acquiring responsibility for local commissioning of health services and their integration with other local services such as social care and housing improvements.

THE REVIVAL OF PUBLIC HEALTH

In the nineteenth century, large sections of the population of towns and cities in Britain did not have access to clean water, clean air, safe food, sewerage systems, or services for collecting and disposing of waste. The effect on health was severe. A person born in 1841 could expect to live a mere 40 years.

By 1989, the life expectancy of men had risen to 73 and of women to 78. The largest decline in premature mortality (death before the age of 65) occurred during the period 1900 to 1939. This was largely a result of basic reforms which reduced the number of deaths from infection. Medical officers of health (MoHs) were particularly important in pressing for housing improvements. Byrne *et al.* (1986: 19) cite examples from Gateshead, illustrating the significance of slum clearance to the fight against tuberculosis (TB) in successive MoH reports from 1912 through to 1933:

to cure cases in the small houses is impossible, to expect to prevent infection is equally impossible, unless the houses are properly constructed, and provision made for efficient ventilation. (1912)

92.3 per cent of cases of TB are to be found in houses of four or less rooms. (1930)

One hopes that the big drive which is being made in slum clearance will bear fruit in the not too distant future in a reduction of the incidence of, and mortality from, Tuberculosis. (1933)

Big improvements in death-rates continued after the Second World War. Better nutrition, clean water, better housing and rising incomes enabled people to resist infection or reduced their exposure. These improvements were also partially due to better medical care, such as immunisation and the use of antibiotics after the mid-1930s, coupled with improvements in maternity and child health services. Many of these services were provided by local authorities before the reorganisation of health services in 1974 when the NHS took over the health services previously run by local authorities.

Replacing or upgrading unhealthy housing and providing clean water were *preventive* actions which had a dramatic effect in improving public health. Preventive health services, however, were already the poor relation of curative medicine when the National Health Service was established in 1948. As a result, the NHS grew as a sickness service rather than a health service.

Since the inception of the NHS, health policy in the UK has been dominated by issues concerning the funding and organisation of medical care. These issues have occupied a central position in the politics of public policy because the NHS touches everyone's lives, it has a million employees and costs the taxpayer some £35 billion per annum.

When the NHS was established, it was thought that its cost would begin to decline as the population grew more healthy as a result of free access to health care. In fact, the cost of the NHS grew considerably and continues to do so. Today, current revenue expenditure on health-care services in the UK is in the region of £4 million *per hour.* Despite this, as was noted in Chapter 4, the UK ranks 22nd out of 23 major industrial countries in the Organisation for Economic Co-operation and Development according to the proportion of gross domestic product spent on health services (*The Guardian*, 25 March 1992). Many people have secured private health insurance as an alternative to the NHS, with one in six UK households now subscribing to this sector (*The Times*, 25 August 1992).

The demands on hospitals, general practitioners and community health services continue to grow. As a result, two issues are now centre-stage in health policy: how to ration health care and decide priorities for treatment, and how to prevent disease, especially the modern-day epidemics of coronary heart disease and cancers. In 1992 the government responded with a White Paper, *The Health of the Nation* (Department of Health, 1992). This re-emphasised disease prevention and health promotion, linking a new national health strategy to a series of targets for improved health (see pp. 267–71).

The rediscovery of public health by a Conservative government seems ironic in view of the right's ideology of rolling back the state. The public health reforms of the nineteenth century introduced by the Victorians, 'in many ways ... represented a bridgehead for the modern, interventionist state' (Baggott, 1991: 192). The new public health has a focus on specific modern-day health problems. These include AIDS, coronary heart disease, drug and alcohol problems, smoking-related disease, accidents, cervical and breast cancer, food poisoning and unhealthy diet, and health problems caused by pollution and deprivation. A clue to the contemporary focus on public health is that these problems are growing; they are difficult, expensive or impossible to cure; and they are largely preventable. The new public health places considerable emphasis on lifestyle factors: educating, informing and persuading individuals to adopt healthier lifestyles. Given the high and growing cost of medical treatment, this approach accords with both promoting individual responsibility and seeking economies in public spending.

The public health function was strengthened to support the health care purchasing function introduced by the 'purchaser–provider' split in the NHS in April 1991. Health authorities and GP fundholders purchase health care from hospitals and community health services in an 'internal market' (see Chapter 3). District health authorities must employ a director of public health, a post equivalent to the old medical officer of health in local authorities which was abolished as part of the 1974 reorganisation of health services. Contracts with providers are expected to be informed by population need assessments carried out by public health specialists in district health authorities, led by the director of public health. The director of public health also produces a public annual report which describes the health of the district's population and identifies health issues which require action. The Department of Health requires that public health considerations must inform all NHS activities. This includes district health authorities entering into 'healthy alliances' with family health service authorities, local authorities and other agencies to promote health. All parts of the NHS are required, 'to ensure that they are able to discharge their responsibilities to maintain and improve the health of the population including arrangements for the control of communicable disease and infection and for dealing with the health aspects of non-communicable environmental hazards' (NHS Management Executive, 1993).

There is a political consensus about the need to diversify health policy away from a concentration on curative medical treatment towards a concern with disease prevention and health promotion as well. However, there are major disagreements about approach, with Conservative government policy criticised for failing to regulate more strongly the smoking, alcoholic drinks and food industries, and for down-playing the effects of deprivation and inequality on health. The Labour Party places emphasis on reducing health inequalities and together with the Liberal Democrats has a broader public health focus than the Conservatives. The Greens' emphasis on removing the social causes of ill health to reduce stress, deprivation and harmful lifestyles is very much in tune with the public health approach.

Today's epidemics of heart disease, cancers and AIDS are strongly linked to unhealthy lifestyles such as smoking, drinking, lack of exercise, eating fatty foods and unsafe sex. Particular attention has focused on smoking. During the 1990s, smoking in developed countries will account for 30 per cent of premature deaths among people aged between 35 and 69 years.

There is a growing awareness that health, and health-related behaviour, are affected by stress, poverty, isolation and a wide range of environmental factors. People's lifestyles are to a large extent determined by their jobs, income and environment, which in turn have effects on their health. For example, healthy diets with less fat, less sugar, and more fresh fruit and vegetables are not affordable for people who are unemployed or on very low incomes (Hall, 1990). Jenkins (1991) reports extensive nutritional deficiencies concentrating among people who are ill, unemployed, receiving benefits or in social classes III, IV or V (semi-skilled and unskilled workers and their families). Deprived areas often do not provide opportunities for exercise that are local and cheap (Vines, 1989). Smoking is increasingly concentrated among women in social classes III, IV and V, and among parents in poverty; research has revealed how smoking helps mothers caring for the family on a low income cope with stress (Graham, 1984).

Accidents – which are the third largest cause of death before age 65 after cardiovascular diseases and cancers – are also strongly correlated with social class (Quick, 1991). Men aged 20–64 who are unskilled have a chance of dying from fire and flames eight times that of men of the same age who are in a professional occupation. The risk of death from suffocation is five times greater and from all other accidents (excluding drivers and pedestrians) six times greater. Unskilled men are slightly less likely to be killed when driving because they drive less, but are eleven times more likely to be killed as pedestrians. The picture is similar for women, although rates are lower and class differences are less sharp. The class gradient for child accidents is stronger than for any other cause of death; the death-rate from all causes for children from social class V backgrounds is twice that of children from professional home backgrounds but three times for accidents. The differences are especially marked for pedestrian deaths and deaths from fires.

The vast majority of diseases, injuries and causes of premature death are thus strongly associated with the incidence of social deprivation. This is as true of diseases that are linked to lifestyle and behavioural factors such as smoking as it is of diseases that are not (Quick and Wilkinson, 1991). Unskilled men have 2.5 times and unskilled women nearly double the standardised mortality rates of professional men and women. A social class gradient exists for both adult mortality and still births and deaths in childhood, and this gradient appears to have been steepening (Delamothe, 1991). Reducing risk factors by encouraging more healthy lifestyles will only bring about a minor reduction in the underlying physiological causes of ill-health. Health inequalities are related to multiple socio-economic influences, although relatively little is known about this. Indeed, tackling single factors, such as health education campaigns against smoking or raising tobacco tax, may have unintended negative effects on health inequalities. Quick and Wilkinson give the following illustration:

> It is often said that health education campaigns widen class differentials in smoking and that it would be better to increase the tobacco tax. But it has been calculated that smokers among the poorest twenty percent of households already spend nine percent of their disposable incomes just on tobacco tax. Paying additional tax may well add to the health problems of those who are not deterred from smoking. A realistic policy for reducing inequalities in health must then address itself to the underlying socioeconomic inequalities.
>
> (Quick and Wilkinson, 1991: 24)

Making healthy choices can be very difficult for people who have to live on very low incomes and have to cope with the stress and loss of self-esteem which this can cause. Health education messages have least impact among people most at risk of having to live unhealthy lifestyles (Hogg, 1991). Urban policy has a role in making healthy choices easier to make. The following are some examples of this approach:

1. Support for child care has benefits for women under stress, and can help reduce the use of tranquillisers, excessive smoking and generally improve their mental health.
2. Provision of school breakfasts in areas with many low income families can help improve children's health and resistance to infection.
3. Food co-operatives are a means of improving the quantity and quality of food available in deprived areas.

These are non-health care services which nevertheless produce a 'health gain', the term used to describe improvements in the health status of a population, such as greater life expectancy, reductions in premature death and improvements in quality of life. Although the NHS is a highly valued service because of its role in treatment and cure, it is only one of many factors which influence the health status of the population. It also has a marginal effect on health inequalities; only small differences in health result from differences in health care in the UK (Quick and Wilkinson, 1991). In terms of prevention, local authorities have a more significant role than the NHS because of their housing, social services, environmental and consumer protection, and education functions. Local authorities already work closely with health services in the areas of social care, environmental health and health promotion, and this collaboration has grown in recent years. The purity of water, adequacy of drainage and quality of housing are no less important to health now than they were in the nineteenth century. When these fail, health is at risk. There continues to be a need to monitor, regulate and control poor housing and infrastructure, pollution, workplace health and environmental problems. Local government services remain central in these tasks.

Local authorities, however, have no significant influence over people's incomes, and it is income differences that exert the most powerful influence on health. As has been noted, there is considerable evidence that variations in morbidity and mortality are caused by deprivation. There is also evidence that income inequality itself causes significant differences in health as a result of its effects on levels of stress, insecurity, self-esteem and social and material exclusion from 'normal life' (Wilkinson, 1992). It appears that the greater the income differentials in a society, the more strain it is likely to place on individuals within it and the more serious its social problems are likely to be.

THE ROLE OF LOCAL GOVERNMENT

Despite the persistence of socio-economic causes of ill-health, local authorities can ameliorate them by providing opportunities to lead a healthier lifestyle and improving the physical and social environment through their services and regulatory work.

Local authorities make an essential contribution to public health through a range of services: environmental protection and trading standards, accident prevention, transport, social services, housing, planning, leisure services and recreation facilities, education and youth services, waste management and emergency services. In two key services, education and housing, local authority influence has been reduced, but one effect of this has been for local government services to become more concerned with particularly vulnerable people such as children with special needs and people with acute housing difficulties. Local authority services have vital implications for the health of these groups.

There are many examples of specific health initiatives by local authorities. These include road safety and accident prevention work; educating young people in schools and youth projects about HIV, AIDS and sexual health; promoting leisure services to encourage physical activity; promoting health in the workplace and developing smoking policies; joint commissioning of mental health services between social services departments and health authorities; alcohol advisory groups; 'look after your heart' campaigns and promoting healthy eating in schools. Local authorities are also large employers and have important contributions to make through employer-based initiatives and occupational health schemes.

There are four major areas in which health is a concern of local government services: environmental and public protection, housing and land-use planning, health promotion, and social services.

Environmental and public protection

These are local government services which safeguard the health of the public by monitoring environmental hazards and taking enforcement action. They are:

1. Control of pollution from domestic emissions and emissions from commercial and industrial processes not covered by Her Majesty's Inspectorate of Pollution.
2. The enforcement of health and safety regulations for service industries and the commercial sector (other industries are covered by the national Health and Safety Executive).
3. Food safety and hygiene.
4. Control of vermin.
5. Control of neighbourhood noise.
6. Enforcement of the Consumer Protection Act which provides a general safety requirement for consumer goods.

Unfortunately, local authority trading standards and environmental health departments are often under-funded and under-staffed. It is possible that the need to meet national and European standards will see greater centralisation of these functions in central government agencies. This would be a major loss to local authorities, which argue that they are well-placed to identify the effects of pollution and that their powers should be increased.

It is the Clean Air Acts which have enabled local government to realise its major achievements in pollution control. By designating urban areas as Smoke Control Areas, significant reductions in smoke and sulphur dioxide emissions have been brought about. The resulting improvements in air quality have made towns and cities much more pleasant places to live, and have resulted in health gains such as the reduction of chronic chest disease.

Table 10.1 shows the impact of Smoke Control Areas in Gateshead during the 1980s. Smoke and sulphur dioxide pollution were reduced dramatically, so that levels in Gateshead are now at or below the national average and well within EEC directive limits.

Trading standards are another monitoring responsibility by which local authorities protect the public from dangers to their health. This involves enforcing legal requirements regarding the sale of goods. Local authorities can monitor the safety and mis-description of goods sold to local residents. They have powers to restrict or prohibit the sale of products and to prosecute traders who sell dangerous items, such as unsafe toys or upholstered furniture which does not meet fire safety requirements. Trading standards departments also deal with complaints from the public and monitor cars for sale on garage forecourts, taking action where vehicles are unroadworthy and therefore dangerous. Other responsibilities of trading standards officers are enforcing legislation on the sale of cigarettes to children under 16 and

Table 10.1 Smoke and sulphur dioxide pollution 1970–1990 (microgrammes per cubic metre)

	Gateshead		*National average*	
Year	*Smoke*	SO_2	*Smoke*	SO_2
1979	104	55	25	57
1980	31	63	19	48
1981	28	42	23	50
1982	25	31	17	40
1983	20	33	18	39
1984	28	40	15	38
1985	23	36	15	36
1986	40	36	17	35
1987	28	22	14	35
1988	27	39	16	32
1989	19	21	15	30
1990	8	13	14	29

Source: Henley, 1991.

advising retailers about the supply of products which can be used for solvent abuse.

Local authorities have many licensing functions which are used to protect the public from danger and to prevent accidents. Licensing is a form of authorisation which allows risks to the public to be assessed and any necessary controls to be imposed. These functions range from licensing major sporting and entertainment venues to ensuring the safe storage and sale of explosives and dangerous substances. Local authorities also license landfill sites and other waste facilities, ensuring that they do not cause water pollution or a health risk (although waste regulation is likely to be transferred to a national Environment Agency in 1995).

Over one million premises catering for seven million workers are subject to inspection by local authority environmental health officers under the Health and Safety at Work Act of 1974. These inspections are aimed at ensuring that health, safety and welfare standards are maintained. Accidents and dangerous occurrences are also investigated.

An increasingly competitive business environment has caused concern about the effects of cutting overheads and boosting productivity on the health and safety of employees. Health promotion, however, can have beneficial effects for employers by reducing sickness absence rates or ill-health pension costs. Health in the workplace is directly relevant to local authorities themselves. Many seek to set a good example to other employers by employing specialist medical advisers, adopting policies on smoking and alcohol, and encouraging women to take advantage of cancer screening services.

There has been an increasing degree of involvement of local authorities in community safety and crime prevention. Local crime prevention strategies often involve the local council, the police and the probation service. Some councils have undertaken special initiatives to address domestic violence, racial harassment, burglaries and the safety of the built environment.

The reduction of environmental hazards is a key function of local authority traffic engineers and road safety officers. This work aims to promote safe movement and reduce serious accidents to motorists and pedestrians. The implementation of traffic-calming measures, the installation of traffic lights and pedestrian crossings, and road safety training programmes all make important contributions to accident prevention.

Accident prevention generally is an important area for 'healthy alliances' because of its multidisciplinary nature. Police, road safety officers, teachers, home safety officers, health education officers, trading standards officers as well as members of the health professions, particularly health visitors, may all be involved at a local level. Local accident prevention groups are common, involving health authority and local authority officers as well as voluntary groups. Community-based projects such as home safety equipment loan schemes have proved to be effective (accidents in and around the home account for nearly 40 per cent of all fatal accidents).

Traffic accidents are one of the most frequent causes of death and injury for young people. Local authorities have a role in creating better public awareness, education and training, and engineering measures to reduce the number or severity of accidents. They are often promoters of public transport by, for example, eliminating or discouraging car use in town centres and other sensitive locations. The likelihood of a road accident depends on the length of journey and the accident risk per mile. The latter varies markedly between different modes of travel, with buses by far the safest type of transport (Hannah, 1991).

The emphasis on private road transport in national government policies has contributed heavily to air pollution, with resulting increases in asthma, bronchitis, emphysema and angina, as well as rising concern about the effects of vehicle exhausts. Reducing the number of cars and lorries would improve public health through fewer road accidents, less pollution, less stress and more cycling and walking.

The late 1980s saw Britain struck by a series of food crises. Salmonella in eggs, listeria and bovine spongiform encephalopathy (BSE – 'mad cow disease') attracted high-profile publicity and saw the rapid introduction of legislation in the form of the 1990 Food Safety Act (also prompted by a European Community food directive). The act requires food outlets to be registered, a weaker measure than the licensing argued for by health and consumer groups (Baggott, 1991). It introduced a stronger regulatory system and extra funds were provided for enforcement.

Food poisoning can be fatal, particularly among older people. The cause is often a breakdown of hygiene standards in the food business. Examples include food not being stored under refrigeration, the cross-contamination from raw food to cooked food and bad personal habits. Food hygiene regulations have been introduced to prevent these problems. Local authorities are responsible for ensuring that food premises, from small corner shops and hot-dog sellers to large food manufacturing plants and hospital kitchens, comply with the legal standards. Environmental health officers inspect these premises to check that hygiene standards are in order and to provide advice to food traders.

Enforcement of hygiene standards has to be complemented by training food handlers in good food hygiene standards. Food hygiene training courses are run by local authorities. Monitoring is also carried out, including regular bacteriological and chemical sampling of food and water.

About half of food poisoning cases are caused by bad practices in the home. This problem is mainly addressed through food hygiene campaigns.

Pest control can also be considered under the heading of environmental health. Rats, mice, cockroaches, fleas, flies and bed bugs can cause damage to health as well as nuisance. These pests can spread disease, cause damage to property and annoyance to people. Local authorities have teams of trained pest control officers. They carry out the destruction of pests and advise on prevention.

Finally, emergency planning is a local authority responsibility. Plans have to be in place to cope with contingencies such as major transport accidents, serious weather conditions, large-scale pollution incidents and other major hazards.

Housing and land-use planning

Good housing standards are essential for healthy living conditions. Local authority programmes of slum clearance, new building, the rehabilitation through grants of older housing, and the provision of basic amenities have brought about substantial improvements in the housing stock.

Local authorities are well-placed to co-ordinate the provision of housing. Increasingly, they have become involved in supporting the needs of particularly vulnerable groups such as young people leaving care, homeless people and women escaping from domestic violence.

Housing policy and practice is where local authorities can have a significant impact on health. There is now substantial evidence of a link between housing conditions and health risks, as summarised in Table 10.2.

The 1855 Nuisances Removal and Diseases Prevention Act established the principle in British public policy of setting a minimum standard to trigger housing intervention. Local authorities could close a house if it was lacking in, 'privy accommodation, means of drainage or ventilation' or if other nuisances were, 'such as to render a house or building, in the judgement

Table 10.2 Health risks and housing defects

Housing defect	*Health risk*
Inadequate heating facilities	Bronchitis, pneumonia, stroke, heart disease, hypothermia, accidents
Damp and mould growth	Respiratory and other diseases
Inadequate ventilation	Respiratory complaints, carbon monoxide poisoning
Lack of hygiene amenities	Infections
Inadequate kitchen facilities	Accidents, food poisoning
Disrepair	Accidents, fire, infections
Structural instability	Accidents
Inadequate lighting	Accidents
Hazardous materials (e.g. asbestos)	Cancer
Overcrowding	Infections, stress
Inadequate means of escape	Injury or death from fire

Source: Audit Commission, 1991a.

of the Justices, unfit for human habitation'. The principle was strengthened by the Artisans and Labourers Dwelling Act of 1868, which gave local authorities, 'provisions for taking down or improving dwellings occupied by working men and their families which are unfit for human habitation'.

The first definition of a minimum standard for fit housing was made in 1919 and, in revised versions, formed the backdrop to large slum clearance programmes in the 1930s and 1960s. The basic elements of the standard have, in fact, changed very little over the century. They are absence of dampness prejudicial to the occupier's health, satisfactory lighting and ventilation, proper drainage and sanitary conveniences, freedom from disrepair, an internal water supply, and suitable facilities for preparing and cooking food. Structural stability and a satisfactory internal arrangement within the dwelling were added in 1954 and 1969 respectively.

Today, housing defects are measured in terms of whether a dwelling lacks basic amenities such as an inside toilet, whether it is in serious disrepair and whether it is unfit for human habitation because of problems such as damp or lack of facilities for preparing and cooking food.

Local councils have powers to tackle defective housing which are implemented by environmental health officers. They include:

1 The closure and demolition of unfit properties where necessary (an uncommon solution since the clearance programmes of the 1970s).
2 Requiring owners of defective properties to carry out repairs and improvements to a minimum standard.

3 Providing means-tested grants to owners and tenants for repairing and improving sub-standard properties.
4 Taking compulsory action to reduce overcrowding.

These powers cannot be used by local councils against their own housing departments. The council housing stock tends not to be associated with the same degree of health risks as housing owned by private landlords, largely because it is generally more modern. However, bad council housing is a cause of health problems and some councils are targeting improvements that can reduce health problems associated with particular parts of their housing stocks (see Chapter 7).

Local authorities can also declare neighbourhood renewal areas (NRAs) where housing conditions are generally dissatisfactory and would be most effectively tackled by comprehensive renewal. A survey carried out to support the declaration of an NRA in Newcastle found that stress was a particular problem among residents. Correlations were found between stress and the existence of damp, draughts, not being able to keep warm and having suffered a burglary (Blackman, Harrington and Keenan, 1993).

The Department of the Environment and the Welsh Office carry out national house condition surveys every five years. However, the sample size of these surveys is not large enough to enable data to be broken down to the level of a district council area. Information on housing conditions at this local level has been patchy. Local councils have recently been given a new statutory duty to consider housing conditions in their districts at least once a year. This necessitates regular sample surveys of the housing stock to identify the scale, location and nature of poor housing. The planning of appropriate housing action also requires information to be collected about tenure and the age and income of local residents.

The availability of resources has been a crucial factor in government decisions about what should constitute a legally enforceable minimum standard of housing. The setting and enforcement of standards has been driven by the money that is available either to replace or improve the housing stock. Moore observes that:

> While conditions have changed, standards and their enforcement have continued to be concerned as much with political, social and economic problems as with the physical needs of the stock. They have tended to rise and fall in line with general economic conditions ... In short, it is difficult, if not unrealistic, to see standards as objective criteria independent of housing policies and resource allocations.
>
> (Moore, 1987: 11)

The same factor has constrained local authorities from taking action against landlords of defective properties because this may involve the authority in having to pay out a large mandatory grant for renovating the property (Audit Commission, 1991a).

The problem of defective housing is concentrated in the private rented sector and especially houses in multiple occupation (HMOs). HMOs are houses divided into bedsits or flatlets, hostels and hotels used for low-cost lodgings. They have been a particular focus of housing action by local authorities. The problem has been reduced dramatically from when this type of accommodation was often overcrowded, in a dangerous condition and insanitary. Today, standards are maintained by regular inspections. In some cities, however, the age and mix of properties, and factors such as large student populations, create continuing problems.

One example is Cardiff where the exploitation of tenants by some private landlords has prompted a major initiative by the City Council. It abandoned a reactive approach based on responding to tenant complaints in favour of a proactive approach based on surveying major high-risk HMOs, together with a cycle of regular inspections, to enforce health and safety standards across the board (Kelly, 1993). Prosecutions, control and possession orders, and withholding housing benefit payments are used to force landlords to improve properties found to be sub-standard. A novel method used by the council is the 'stigma sign', a large poster hung outside the more ramshackle properties proclaiming, 'The landlord has failed to look after this rented property so Cardiff Council is doing the work at the landlord's expense.' However, positive incentives are also important. These include rent guarantees and grants. Private landlords have been supported by the council in organising their own association. A forum which brings together council officers, landlords, tenants and other agencies has direct links into council policy-making. Cardiff's proactive approach is now reflected in national guidance about action on HMOs issued to local authorities by the Department of the Environment (*Inside Housing*, 13 August 1993).

Some local authorities have adopted a targeted approach to monitoring and enforcing housing standards generally, concentrating their environmental health officer teams on priority areas. This can also be a basis for inter-agency collaboration. For example, the local health authority can target its health promotion activities on such areas, the local council can target its monitoring, inspection, enforcement and cleansing activities, and community groups can be involved in joint planning with statutory agencies.

As well as housing action, local authorities produce land-use plans which include policies to safeguard areas of high environmental quality and to promote the upgrading of areas with poor environments. The protection of open spaces, improving access to public transport and promoting employment and local services are important land-use planning functions. Development control and building control are regulatory functions which restrict new developments which might harm the amenities and lifestyles of residents. Planning and design can be used to reduce the risk of accidents and to reduce fear and stress associated with poorly lit streets and areas perceived to be dangerous.

Health promotion

Leisure departments are major multi-disciplinary leisure organisations in the larger urban local authorities and key features of many other district councils. Their origin is in nineteenth-century public health legislation which saw the creation of parks and open spaces. Subsequently, public sector leisure provision expanded to include swimming pools, sports centres and children's play provision, as well as facilities such as libraries and art galleries.

Many of these services are important in providing opportunities for exercise, relieving stress and personal development. However, recent Conservative government policy towards local authority leisure provision has introduced a greater commercialism, including the compulsory competitive tendering of leisure services. This is endangering the egalitarian and collectivist purposes of direct provision by local councils. Despite this, targeting healthy leisure opportunities where needs are greatest remains an important objective of many authorities. Methods include prioritising capital and revenue spending geographically, and schemes such as 'leisure cards' which target subsidies on individuals who are unemployed or have low incomes.

The promotion of health awareness and physical exercise in schools is an important element of the National Curriculum. School and education welfare staff are key players in the child protection service, and schools work closely with GPs and community health services to monitor the physical health and development of all schoolchildren.

The commitment of central government to health promotion in schools was put into question by its decision to end special funding from March 1993 to local education authorities for the support of health education co-ordinators. The delegation of school management to governors and head teachers, and the possibility for schools to leave the control of their local education authorities to become grant-maintained, have also made it difficult for local authorities to implement policies for health promotion in schools.

Health and social services

Local authorities provide a large range of personal social services. For children, these include family support, placements for children in need, and the registration and inspection of day care. For adults, services include home care, residential homes, aids and adaptations, support for people with mental health problems, and services for people with drug or alcohol problems and for people affected by HIV. Increasingly, local authority social services departments are purchasing services from providers in the private and voluntary sectors rather than providing services directly.

The NHS and Community Care Act of 1990 aims for a 'seamless service' which bridges health and social care for adults with medical and social care needs. This is particularly the case when a person discharged from hospital needs a package of social care at home, although GPs also have an important role in making referrals to social services departments of patients who are living in the community and may need social care. Present policy aims to transfer prime responsibility for social care to local authorities, leaving health services to devote their main energies to treatment and assisting with physical and psychological rehabilitation.

The new policy was fully implemented in April 1993. It has a strong focus on assessments of need. These assessments are used to make decisions with users and carers about appropriate services. Social services departments place contracts with providers of services to meet the needs they assess. Services are rationed by applying eligibility criteria before they can be accessed. Many local authorities appoint 'care managers' to plan and monitor a person's package of care. The local authority pays for the care provided but will charge individuals according to their means, whilst health services remain free to the user. Whilst these new procedures have made the assessment of need and provision of services more systematic, and with greater accountability to users and carers, they are time-consuming and entail extra paperwork. Many users will not receive any improvement in the care they actually receive.

Although the community care changes seek to bridge health and social care needs, these are still separate responsibilities for the NHS and local authorities. When a general practitioner (GP), for example, thinks that a patient has a social-care need, the patient is referred to their local authority's social services department. A social worker (usually) makes an assessment of the person's needs, involving the person as fully as possible in the assessment. The social worker must also involve other appropriate professionals, including the GP. In particular, medical or nursing advice must be sought if admission to a residential or nursing home is being considered. Permission from the health authority is required before placing a person in a nursing home. Wherever possible, services will be provided which enable the person to carry on living at home.

Local authorities have a lead role in developing community care with partners in the NHS, the voluntary sector and the private social care sector. As well as assessing individual needs and providing packages of services round these individual needs, local authorities are responsible for producing an annual *community care plan* for the development of community care services. These plans are expected to contain the following elements:

1 Clarification of values and principles.
2 Definition of objectives based on these principles, e.g. enabling people to live as normal a life as possible in their own home.

3 Identification of types and levels of need from assessments, consultation, surveys and census data.
4 Auditing supply – establishing the extent and nature of provision by the statutory, voluntary and private sectors for different categories of need.
5 Developing purchasing strategies and frameworks which seek to match supply to needs within available resources, including prioritising, setting targets, identifying funding responsibilities and specifying contracts.
6 Defining agency responsibilities, e.g. between health and social services, or social services and the housing department.
7 Integrating plans with financial planning, e.g. allocating budgets according to need assessments.
8 Reviewing performance against targets and contract specifications. Reviewing values and objectives against experience of service provision and changes in levels and types of need.

As noted above, medical care remains the province of the NHS, but the new community care policy seeks to co-ordinate the provision of services for social care and medical care when, as is often the case, an individual has need for both. This has been done by local councils and health authorities adopting formal agreements about assessment, care management and discharge from hospital. There is an increasing trend for health authorities and local authorities to undertake joint commissioning of care services.

One of the major pressures for joint planning and commissioning is trends in health care. Hospital beds are being used more efficiently. More operations are performed as day surgery, which does not need an overnight stay in hospital, and stays for people with acute illness are shorter as a result of rapid advances in techniques and the increased effectiveness of treatments. In addition, more treatments can now be undertaken outside hospital in clinics, by GPs and at home. Each year the health service treats more patients, and demand continues to rise.

These trends mean that 'bed blockage' is a real issue for hospitals which have patients who are no longer able to benefit from hospital care but whose discharge home is prevented by insufficient support being available in the community. This support includes informal carers such as relatives and friends, community health services provided by nurses, doctors and other health-care professionals such as chiropodists, and social care services such as home helps, residential or nursing homes, and the provision of aids or adaptations.

Many patients and their families prefer care services to be delivered to the home, or provided as near to home as possible. There is evidence that people recover well with nursing and therapy at home rather than having to stay in hospital. New technologies and the provision of intensive support make sophisticated home care possible, such as intravenous chemotherapy and home renal dialysis.

The 1990 NHS and Community Care Act places great emphasis on the services provided by primary health care teams and social services staff. The balance of spending between primary care (GPs and services such as health visitors) and acute hospital services (secondary and tertiary care) is being shifted to the former. District health authorities and fundholding general practices must in their purchasing plans make provision for increasing numbers of people who, as a result both of the new arrangements for community care and changes in the pattern of acute care, will be looked after in the community. Indeed, the main aim of community care is to enable people with social or health care needs to live as normal a life as possible in their own homes or in a homely environment in the community. This aim is widely accepted, although there is considerable controversy about whether funding is sufficient to provide the support that people need to live in the community (see Chapter 4).

Social services budgets for community care are cash-limited. There is a danger that by targeting services on those in greatest need, resources to meet the needs of people requiring minor help will be scarce. This undermines preventive action. For example, minor home adaptation may prevent the frail and elderly falling and enable them to remain independent at home. But demand for adaptations far exceeds supply and these services may not be available until after the person has descended into a crisis and is high need, such as suffering a hip fracture. Cash-limited budgets may also put informal carers, who are often close female relatives, under additional pressure to care for very old, ill or disabled dependants at home.

Social services departments are now purchasers of social care in the same way that district health authorities and fundholding GPs are purchasers of health care from suppliers such as hospitals, community services and ambulance services. Social care is now provided in a 'mixed economy': providers can be social services departments, voluntary organisations or private companies. In the NHS there is an 'internal market', with hospitals in competition with each other for contracts from health authorities and fundholding GPs. Both are 'quasi-markets' in the sense that social services departments and health authorities purchase services on behalf of individual consumers; they are not normal markets in which consumer choice is reflected directly in individuals' decisions about what to buy.

The adoption of an explicit purchaser/provider split in both health and social services has required that both types of authority assess the needs of local populations, prepare plans to meet these needs, specify contracts with providers, and monitor contracts to ensure accountability and quality. Health and social services are performing the same strategic tasks and providing services that are frequently complementary.

THE HEALTH OF THE NATION

In July 1992, the British government published its national long-term health strategy in a White Paper, *The Health of the Nation* (Department of Health, 1992). The strategy reorientates health policy away from curative health services and towards disease prevention and health promotion. The provision of health services is only one dimension of an overall public policy approach which has the following elements:

1. Consideration of the health dimension when developing any public policy. Examples include housing improvements (which have been shown to reduce respiratory and mental conditions), safe building designs and road improvements to reduce accidents.
2. Physical environments conducive to health. As well as approaches such as no smoking policies and accident prevention, targets are incorporated from the White Paper *This Common Inheritance* for air and water quality and ultraviolet radiation exposure.
3. Informing people about healthy lifestyles. Examples include health promotion in schools and healthy eating campaigns.
4. High-quality health services. There is an overall commitment to the NHS. NHS services for mental illness are identified as an area needing continuing development.

The overall goals of the strategy are:

1. 'Adding years to life': increasing life expectancy.
2. 'Adding life to years': reducing ill-health and its effects.

The White Paper sets out five priority areas for action, further elaborated in a series of handbooks issued to health authorities and local authorities. Each priority area has a small number of targets against which success or failure can be measured. For example, targets for coronary heart disease include reducing the death-rate in people aged under 65 years old by at least 40 per cent between 1990 and 2000, and in people aged 65–74 by at least 30 per cent over the same period. Targets for accidents include reducing the death-rate among children aged under 15 by at least 33 per cent between 1990 and 2005, and among young people aged 15–24 by at least 25 per cent over the same period.

Approaches to meeting these targets are set out in the strategy. For example, for coronary heart disease the approach is stated as stopping smoking, reducing consumption of fatty acids, alcohol and salt, and increasing physical activity. The influence of socio-economic factors, circumstances in early life and stress are recognised but not fully addressed by the strategy because they are 'less well understood'. A large research and monitoring programme supports the strategy, including a programme of national health surveys. Implementation is overseen by a ministerial level committee on which eleven government departments are represented. In England, three working groups support the committee: the wider health working group is chaired by the minister of health; the health priorities working group is chaired by the government's chief medical officer, and the working group on implementation in the NHS is chaired by the chief executive of the NHS Management Executive.

The 'Health of the Nation' priority areas are considered below with the addition of comment about the role of local government. This comment draws on the circular of guidance for local authorities issued by the three local authority associations in England (ACC/ADC/AMA, 1993).

Coronary heart disease and stroke

Major lifestyle and risk factors are targeted for reduction, including smoking, diet, obesity, blood pressure and alcohol consumption. The White Paper does not propose a ban on cigarette advertising; such a ban is supported by the local authority associations.

Local authorities have the following inputs to make to this priority area:

1. Healthy diet and nutrition (schools, residential homes, meals on wheels, promotion of heartbeat-type award schemes in cafes and restaurants, safeguarding food quality).
2. Workforce alcohol policies, services for alcohol misusers and their families, education through schools and youth services, drink–drive campaigns.
3. Smoking policies (for councils' own premises and premises subject to health and safety enforcement, licensing and food hygiene regulations).
4. Enforcement of the Children and Young Persons (Protection from Tobacco) Act 1991.
5. Physical activity – providing opportunities for people to keep fit and healthy through exercise, development of local strategies for sport and recreation, and physical education in schools.
6. Generally raising awareness and providing education on smoking, diet, alcohol and physical activity.

This area has considerable scope for healthy alliance work between local authorities, district health authorities, family health service authorities, the voluntary sector, the private sector and the media.

Cancers

The main focus of *The Health of the Nation* is on reducing smoking, the major risk factor for lung cancer.

The local authority associations advocate that the large majority of employees should be covered by a no-smoking policy by 1995. The National Curriculum in schools requires children to be taught about the harmful effects of smoking. Local authorities can also discourage smoking through their health and safety enforcement function, pollution control and smoking policies to create a healthy environment for the public and staff.

Mental health

The main target is stated as improving significantly the health and social functioning of people with mental illness.

Local authority social services departments have a key role in co-ordinating community care. District health authorities are the lead agencies for specialist mental health services. Recent years have seen a programme of hospital closures and the development of alternative community-based mental health services. Although community-based care is seen as more desirable, the adequacy of support in the community for people with mental health problems has often been poor and is only slowly improving.

Several areas are highlighted for consideration in developing local strategies:

1. Assessment of baseline information and implementation of effective mental health information systems to assist in developing services and monitoring progress towards targets.
2. A systematic approach to needs assessment and reviewing options for intervention.
3. Local target setting.
4. The development of effective joint planning and purchasing with the NHS to ensure a smooth transition to care in the community.
5. Building broad alliances for mental health promotion (including schools, local media and workplaces).

Resettlement planning, putting in place local mental health services including supported housing, and integration of health and social care demand a multi-agency approach to mental health. *The Health of the Nation* handbook recommends that three- to five-year multi-agency mental health strategies are developed in local areas (Department of Health, 1993b). Positive action to provide employment opportunities for people with learning difficulties, and specialist help and social care for offenders with mental health problems, are also recommended.

Social services, education and youth services, and housing services have direct roles in helping people with mental health problems. Local authorities can also promote mental health by ensuring accessible and affordable recreation facilities, safe environments for children to play, and community facilities for unemployed people, women at home, older people and people with chronic sickness. Tackling crime has become an important aspect of action to improve mental health and can be addressed through housing improvements as well as police initiatives.

HIV/AIDS and sexual health

HIV and AIDS are a major challenge for public health strategies. Prevention depends on promoting sexual health, such as the use of condoms to prevent sexual transmission, and tackling drug misuse (the sharing of contaminated injecting equipment transmits HIV). Local authorities are responsible for providing community care for people who are HIV positive.

A collaborative approach between local authorities, health authorities and voluntary organisations is promoted by *The Health of the Nation*, working within agreed strategies for local sexual health and drug misusing. School-based education is particularly important.

The following ways of strengthening corporate responses are suggested in the Department of Health (1993c) handbook *HIV/AIDS and Sexual Health*:

1 Developing strategic plans for sexual health covering health education and promotion, provision of appropriate services, research and setting targets.
2 Encouraging the development, implementation and review of sex education policies by school governors.
3 Ensuring that health and social services reflect the needs of people with HIV or AIDS.
4 Developing good practice in prevention and service delivery, and sharing experience through networks. Local authorities, for example, can introduce HIV awareness into a range of services such as leisure and recreation, housing, tenants' associations, environmental health, youth services and community education.
5 Purchasing and financing various services to meet practical needs and particular requirements, including counselling of people known to be infected, their partners, families and carers.
6 Developing appropriate HIV training initiatives for staff.
7 Helping to foster better links between statutory and non-statutory services, including joint commissioning between local authorities and health authorities.

Drug misuse is addressed by *The Health of the Nation* in terms of reducing the incidence of HIV transmission. In the mid-1980s local education authorities could bid for an education support grant (ESG) to fund work co-ordinating drugs education in their areas for an initial period of two years. The Department for Education extended ESG for a further two years with an expanded programme including solvents, tobacco and alcohol. In 1990 the ESG for drugs education was replaced with funding for a broader health education programme, including HIV/AIDS. However, in March 1993 this special government funding for health education by local authorities was terminated. Although it was meant to be temporary 'pump priming' support, many local authorities found themselves no longer able to continue the health co-ordinator posts that had been supported with this funding. This has intensified the problem of drugs education in schools being squeezed because it is non-statutory and schools are under pressure to concentrate on the statutory National Curriculum.

Accidents

Local authority responsibilities are wide-ranging in this area. They include trading standards, traffic management and road improvements, traffic calming, land-use planning, safe play and leisure facilities, environmental health, waste disposal, care services and aids/adaptations which enable elderly or disabled people to live at home, education, the design and improvement of safety measures in council housing, fire services, and representation on police authorities.

Membership of local accident prevention groups will include local authority representatives as core members, together with NHS representatives and members drawn from voluntary and independent bodies such as youth organisations, community health councils or the St John Ambulance service.

The national Health and Safety Executive is responsible for regulating health and safety standards in the manufacturing, construction and chemicals industries, whilst local authorities are responsible for shops, offices and warehouses.

Local authority associations have criticised the White Paper's emphasis on reducing death-rates for accidents, arguing that the target should relate to all injuries from accidents that require medical or clinical attention.

HEALTHY ALLIANCES

One of the first local joint health strategies developed in direct response to the *Health of the Nation* White Paper has been the Joint Health Strategy for the London Borough of Brent (Brent Council, 1993). This is a joint initiative between Brent Council, the Brent and Harrow Commissioning Agency (the local health authority) and the Brent and Harrow Family Health Services Authority (responsible for family doctors, dentists and pharmacists).

The strategy reflects the new emphasis on *health gain*: 'adding years to life' (through an increase in life expectancy and a reduction in premature death) and, 'adding life to years' (through increasing years lived free from ill-health, reducing or minimising the adverse effects of illness and disability, promoting healthy lifestyles, physical and social environments, and generally improving the quality of life). It emphasises positive health rather than the more narrow focus of the traditional medical model based on the treatment of illness and the management of health facilities. The Brent approach is based on a social model of health: the health of populations rather than just patients, and the prevention of ill-health rather than just its treatment. The borough's high local levels of deprivation and multi-cultural population are key factors for the strategy to address.

The Brent strategy consists of a series of objectives and a commitment to reviewing the strategy annually. It is based on multi-agency working because it is recognised that no single agency in isolation can tackle the range of social, environmental and behavioural factors which determine health outcomes in the borough. The focus is on health problems in Brent which display marked inequalities or are significantly worse than average.

The first year of the strategy adopted a 'children' theme. Work was concentrated on child accident prevention; school-based health education, family planning and parentcraft education; a healthy eating campaign for pre-school children; safe play areas; a 'Cover Up Kids' campaign about skin cancer; and increases in no-smoking areas. These initiatives are evaluated for their effectiveness.

A Healthy Brent Steering Group brings together senior representatives of Brent Council, the District Health Authority, the Family Health Services Authority and the Community Health Council. Consultation takes place with many other organisations and with the wider public. The service plans and purchasing plans of all key local agencies are encouraged to consider health promotion and the achievement of health gain.

THE HEALTH OF THE NATION: A CRITICAL ASSESSMENT

Although *The Health of the Nation* is the UK's first national long-term health strategy, its approach differs in one significant respect from important reviews of national health carried out in the 1980s. These are *The Black Report* (Department of Health and Social Security, 1980), which presented the findings of the Working Group on Inequalities in Health, and *The Health Divide* (Whitehead, 1987), published by the Health Education Council to update *The Black Report*. The focus of these reports was on health inequalities and their link with social inequalities. Key recommendations were made which needed substantial increases in public expenditure, including raising child benefit, expanding day care for children and expanding local authority housing programmes. The government rejected both reports.

The Health of the Nation makes very little reference to health inequalities and contains no targets to reduce them. Local authorities have criticised the neglect of measures to ameliorate the effects on health of poverty, homelessness, poor housing and environment, and unemployment. The substantial restructurings of the NHS, especially during the 1990s, have been largely irrelevant to the objective of improving health. As the National Children's Bureau commented in 1987:

> Rates of perinatal and infant mortality, child abuse, cigarette smoking and drug addiction all go up in direct proportion in relation to rates of unemployment, homelessness and poverty. Poor perinatal care and immunisation failure are also more common among the most socially disadvantaged groups. We would be failing in our duty if we did not point out that improvements in social conditions would make a greater impact on child health than is likely to occur as a result of any reorganisation of professional work.
>
> (Quoted in ACC/ADC/AMA, 1993: 8)

Children are particularly vulnerable to adverse social conditions and poor physical environments. Particular concern has been expressed about the impact of bed and breakfast accommodation for homeless families on health, development, safety, education and diet (Grieve and Currie, 1990).

Multiple social disadvantage is frequently associated with specific risk factors, such as high smoking rates, alcohol abuse and poor diet. It is unlikely that improvements in lifestyle will be brought about without tackling the underlying social determinants of health, particularly incomes that are too low to support participation in a healthy way of life and which have failed to keep up with rising living standards further up the income scale. Although the real incomes of most poor people in Britain showed a small improvement during the 1980s, their relative position deteriorated because social security benefits did not increase as much as average earnings (*Hansard*, 1992a; 1992b; 1992c). Overall, the incomes of the poorest 20 per cent of households in Britain improved marginally by 4 per cent during the 1980s, but the incomes of the richest 20 per cent rose substantially by 39 per cent. Couples with children and single people who were in the bottom 20 per cent of the income distribution saw their real incomes decline slightly during the 1980s. This widening of income inequality is almost certainly causing a widening in health inequality.

The statistical association between deprivation and health is well-established, but less is known about the causal links. There is a high correlation between mental illness and deprivation for example. One of a number of causal mechanisms indicated by epidemiological research is a link between a higher risk of viral infection for expectant mothers living in poor housing conditions and a higher risk of children developing schizophrenia in later life. The link between housing and respiratory conditions is better understood. Damp is a common problem in sub-standard housing and the fungal growth this can induce produces spores which may enter the respiratory tract, causing bronchial and asthmatic symptoms.

Table 10.3 Calculating the effects of housing and income on health: Gateshead council tenants

Income	*Per cent living in 'good' housing*	*Per cent reported health 'good'*	*Effect of income*	*Effect of housing*
Step 1				
Benefit recipient	66/218 = 30%	103/218 = 47%		
			18%	
Non-benefit recipient	111/130 = 85%	84/130 = 65%		
Step 2				
Benefit recipient	Good housing	39/66 = 59%		17%
	Bad housing	64/152 = 42%	8%	
Non-benefit recipient	Good housing	75/112 = 67%	8%	17%
	Bad housing	9/18 = 50%		

Source: Byrne *et al.*, 1985.

Housing is one of the most neglected areas in *The Health of the Nation.* Poor housing and its related ill effects are seldom improved through individual action alone and require state intervention. Children are particularly vulnerable to the effects of bad housing environments on their health, education and psychological development, and these effects last well into later life. There are also particular problems of affordable and suitable housing for young people (homelessness has increased for this group, partly due to their restricted entitlement to housing benefit) and for older people who may also need social care or adaptations to their home, and may be vulnerable to cold-related illnesses caused by inadequate heating and thermal insulation.

Byrne *et al.* (1985) carried out a causal analysis of the relationships between income, housing and health. Simple cross-tabulations of data were used to test causal models of how housing affects health. The data were from a survey of council tenants in Gateshead. Housing estates were classified as 'good' or 'bad' on the basis of various housing management measures, such as requests for transfer and difficult-to-let indicators. Income was measured in terms of whether or not the household received housing benefit. Respondents reported assessments of their own health to interviewers. The authors considered it important that health was measured on the basis of respondents' own assessments of how they felt, including well-validated diagnostic questions about symptoms.

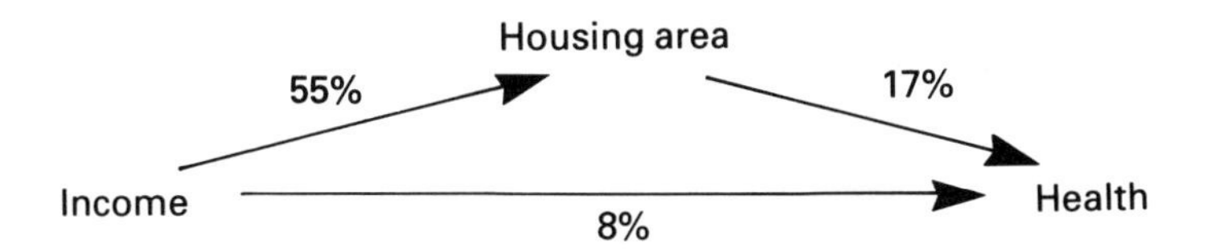

Figure 10.1 A simple model of the effect of housing and income on health: Gateshead council tenants
Source: Byrne *et al.*, 1986.

Table 10.3 shows how the data were analysed to measure the effects of income, housing and health, using the question about general health status as the health indicator. The results from Table 10.3 can be used to construct a simple model as in Figure 4.1.

Table 10.3 and Figure 10.1 can be interpreted as follows. The table shows that income has a simple effect on health of 18 per cent: that is, for households not on benefit, 18 per cent more reported their health as good compared with households on benefit. This direct effect is reduced to 8 per cent when housing area is introduced into the model (Figure 10.1). Income also has a direct effect on housing area of 55 per cent, and housing area has a direct effect on health of 17 per cent. Income also has an indirect effect on health through housing area.

The major policy implication which follows from this study is that health can be improved collectively by improving housing conditions. The authors are critical of strategies which attempt to manage the problem in the worst housing areas, rather than eradicating the cause through demolition. However, they point out that in the past housing policies based on 'sanitary adequacy' produced housing of mediocre quality. Quality 'healthy' housing in the public sector is best guaranteed, they argue, by effective political pressure.

Epidemiological research which seeks to recognise and understand the factors associated with the maintenance of health is essential for making positive health policies. Statistical associations between indicators of ill-health and other variables such as environmental factors can help to develop causal models of health. These can be a basis for policy initiatives even before a full medical understanding of such associations is established. For example, Quick and Wilkinson argue that their evidence on the link between income inequality and inequalities in health justifies income redistribution policies even though our understanding of this relationship is incomplete at present:

> Until our knowledge of the aetiology of the degenerative diseases is much better understood it will remain impossible to account for much of the influence income distribution has on death rates. In the meantime we would do well to remember that rather than detracting

> from the importance we should attach to it, it increases our need to rely on such factors. *That so many causes of death are sensitive to socioeconomic differences may be the most useful thing we known about them.*
>
> (Quick and Wilkinson, 1991: 33–4)

One aspect of socio-economic deprivation which has received a lot of attention is the possible link between unemployment and health. It is known that unemployed people report more illness than people who are in work. This could have one of two explanations. The first could be a tendency for people suffering from ill-health to become unemployed. The second could be that unemployment causes ill-health. A negative correlation between unemployment and health is consistent with both explanations. Bartley (1991) reviews the evidence of studies of redundancy which have investigated whether health deteriorates following job loss – that is, whether unemployment causes ill-health. These studies indicate that stress and ill-health increase with the *anticipation* of redundancy, not as a result of unemployment itself. Other evidence from longitudinal studies suggests that periods of unemployment are not associated with an increase in health problems, but that in the long term people who have experienced unemployment have a higher mortality rate than people who have not.

Bartley seeks to solve these puzzles by thinking theoretically about the effect of social conditions on health. The theory which best fits the evidence about unemployment and health is one which considers the effect of work on people who are vulnerable to unemployment. This includes people with low skills and educational attainment. The work they are able to obtain is often wearing, stressful and temporary. People who are less healthy will be at higher risk of unemployment. People who are healthy will suffer damage to their health. Bartley suggests that it is their experience of insecure and hazardous work which results in a decline in health in the long term. For workers facing redundancy it is probably not anticipation of unemployment which induces stress and ill-health but the fact that firms having to make redundancies have often imposed greater demands on their workers to try and avoid the decline in profits that eventually forces job losses. The overall conclusion is that work insecurity, low pay and hazardous forms of work – which are more widespread when unemployment is high – impose long-term health costs.

Thus, Bartley finds neither the explanation that people suffering from ill-health tend to become unemployed nor that unemployment causes ill-health particularly satisfactory. Rather, general labour market experience seems to be where the explanation lies. The need to address working conditions remains as relevant an issue for urban policy as it was in the nineteenth century.

The work of Quick, Wilkinson and Bartley suggests that both urban policy and policy for the health services are limited in the health improvements they can achieve. Healthy and safe employment together with the reform of taxation and benefits to narrow income inequalities are far more important factors. Indeed, health inequalities appear to be indicators of the overall extent of economic inequality and social division in society.

HEALTH FOR ALL

The Health of the Nation and recent policy documents from the Labour Party and the Liberal Democrats reflect the revival of the public health movement in Britain. In part, this revival can be traced back to 1977 when the World Health Organisation (WHO) launched its 'Health for All' campaign. This promoted targets for national governments to adopt which included the improvement of public health, equity in health provision, community participation and collaboration between agencies. By the year 2000, the main social target is, 'a level of health that will permit all to lead a socially and economically useful life'.

The WHO European Regional Strategy was published in 1981 and similarly promotes the key principles of equity, community participation and inter-sectoral collaboration. Thirty-eight targets were promoted as a framework for local goal setting. In 1987, the WHO created the Healthy Cities Project as a way of putting Health for All into practice at local level. By 1993 there were thirty-five cities participating in the initiative across Europe, including four in the UK (Belfast, Camden in London, Glasgow and Liverpool). Within participating countries, networks of further cities and towns have adopted the Health for All targets. The UK Healthy Cities Network was set up in 1987. In 1991, it was formally reconstituted as the UKHFAN (UK Health for All Network) to include towns and smaller communities. It is funded by membership subscriptions and a grant from the Department of Health.

The Health of the Nation acknowledges the role of Health for All. However, this is marginal to the disease focus of its key areas approach, and the social and environmental prerequisites for good health are not strongly promoted. Despite this, in promoting 'healthy alliances' *The Health of the Nation* provides a new impetus for local partnerships around health objectives.

The UKHFAN has the following aims:

1 Working together towards reducing inequalities in health.
2 Promoting a holistic understanding of health.
3 Encouraging and sharing the experience of collaborative local initiatives.

Birmingham has established a campaign called Healthy Birmingham 2000 (Terry, 1991). Twelve health targets have been identified for the city in the 1990s, including the following examples:

1 To narrow the gap between those with the best and worst health by at least 25 per cent.
2 To ensure people with a disability have the same opportunities as everyone else to lead a fulfilling life.
3 To increase average life expectancy to 75 years.
4 To reduce deaths from accidents by 25 per cent.
5 To reduce significantly the misuse of drugs and other dangerous chemical substances;
6 To involve individuals and communities more actively in the planning of health and other services they need.

This campaign seeks to ensure that all public policies have a health dimension. Among the strategies used to implement Healthy Birmingham 2000 are health action areas in deprived parts of the city. This and other strategies work through links between council departments and with housing associations, voluntary organisations and businesses.

The Health for All and Healthy Cities initiatives have seen a mixture of approaches develop. Some have established practical neighbourhood projects or issue-focused work. Others have worked at an organisational level, promoting training and liaison between agencies to achieve organisational commitment to Health for All objectives.

One of the criteria for cities entering the second five-year phase of the WHO Healthy Cities Project is that they develop city health plans. Several Health for All projects had already begun to move towards this strategic planning stage. Health plans have been used to promote public participation in health strategies and to encourage council departments and other agencies to adopt Health for All principles in their work programmes and strategies.

The Glasgow Healthy Cities Project is among those which have developed a city health plan (Lyon, 1993). The project has the active support of the district and regional councils, the health authority and the two universities. There is also a wide range of involvement from voluntary organisations and community groups. The key aim of the project is to maximise health gain from the expenditure of all the city's key agencies. The plan covers all public services within the city. Three main outcomes are identified:

1 A baseline document which describes the service provision and impact on health of all city council departments and other major agencies. Examples of existing good health practice are also included.
2 A reorientation of organisational cultures towards Health for All objectives.
3 The opportunity for different agencies and the public to work together on health issues.

An example of (2) is a decision by the City Council to establish a health and environment sub-committee. The city health plan is part of the overall work of the project which as well as policy development includes a local action programme, information and training, and liaison with other projects nationally and internationally.

Innovative methods have been used to develop strategic plans. The Rochdale Health for All project, for example, has used arts-based workshops to involve groups of local people in producing a discussion document. This will form the basis of a strategic health action plan. These workshops used collage and writing to explore concerns about health and to assess health needs. Some of the artwork is reproduced in the discussion document. A local community arts project was commissioned to produce a video and exhibition to accompany the launch of the discussion document and encourage community participation.

The Healthy Sheffield project launched its discussion document, *Our City, Our Health*, in 1991. This was followed by a major consultation process. Over the course of a year, 300 trained consultation facilitators were used to encourage active involvement from organisations and the general public. This also included a touring photographic exhibition, extensive media coverage, and supporting newsletters and briefings. Over 1,500 responses were received during the consultation. An analysis of them was fed into the drafting of a policy document, *Our City, Our Health – A Framework for Action*.

More focused than city health plans are 'healthy settings' initiatives. In the UK these include 'health promoting hospitals', the National Healthy Schools Project, a Healthy Workplace initiative and a Healthy Prisons Project. European networks for these initiatives have been established through the World Health Organisation.

CONCLUSIONS

As Baggott (1991: 193) comments, 'Public health is not simply a focus of concern: it is an approach to tackling health and related social problems.' This approach prioritises action at community level across all areas of public policy. Thus, local authorities have a major role to play as partners with health authorities in protecting and promoting the health of individual citizens and whole populations.

Local authorities are strategic bodies. They assess needs across all service areas and seek to influence the actions of different organisations. As the one democratic body at the local level, many councillors feel they have a role in championing the consumer of public services generally. Although the NHS has the central role in *The Health of the Nation*, it cannot represent local communities; health authorities no longer include representative members from local government.

Whilst the Department of Health's encouragement for 'healthy alliances' between health authorities, local councils and voluntary organisations appears to endorse co-operation, there are considerable limits to how far local priorities can drive healthy alliances. The NHS is vertically organised, with policy and priorities decided at the centre and implemented by local health authorities. *The Health of the Nation* is a prime example, setting the targets against which health authorities are expected to perform, largely regardless of local assessments of need or jointly agreed priorities with local councils.

Public health is a prime focus for corporate working across many different services. Although establishing new care services in the community has seen the establishment of joint committees between local councils and health authorities, such inter-agency arrangements are much less common in public health. The Healthy Cities project has been particularly important in stimulating a reassessment of urban policy along public health lines, and promoting joint working in areas such as accident prevention and homelessness. However, healthy public policies should mean that health objectives and performance measures are integral to the business plans of the whole local public sector.

Although the separate control of health and local government services and the degree of central direction in the NHS can be obstacles to health strategy at local level, there remains considerable potential for local councils to develop a more explicit role for themselves in health policy. There is a need for closer management between health, social, housing and environmental services. Developments in this direction have included adopting the same patch areas for delivering services and jointly funding some services. However, as local authorities develop their role in assessing needs and ensuring that services are available without necessarily directly providing them, there is a case for the commissioning role of district health authorities to be transferred to local government. Not only would this introduce local democracy into the health service for the first time, it would also be likely to replace a 'sickness service' with a 'health service' covering medical services, long-term care, preventive strategies, environmental health, housing, social services and health education.

Part IV

ISSUES FOR URBAN POLICY IN THE 1990s

11

ISSUES FOR URBAN POLICY IN THE 1990s

This final chapter reviews the UK's experience of urban policy and prospects for the future. Although the outlook on many fronts is not good, there are some hopeful signs. Future prospects depend on having a clear vision of the aims and objectives of urban policy. These are conventionally regarded to be about meeting 'needs', but the concept of 'need' raises many issues about how decisions are made. The chapter discusses how an urban policy which is democratic, strategic and enables the management of services close to the consumer might be achieved. It makes a case for strengthening urban policy by planning state intervention at a regional level. But in contrast to the regional arms of government as they presently exist, the chapter proposes a democratic model based on regional government. It concludes by returning to the theme of sustainability as a reminder that this must be the fundamental guiding principle of urban policy in the future.

PROSPECTS FOR URBAN POLICY

A Policy Studies Institute report published in 1992 concluded that during fifteen years of urban policy, 'surprisingly little has been achieved' (Willmott and Hutchison, 1992: 82). The UK's first comprehensive urban policy was launched by the 1977 White Paper, *Policy for the Inner Cities*. If the objective of urban policy had been to reduce the gap in well-being between deprived areas and the rest of the country, it had failed:

> in general the gap between conditions and opportunities in deprived areas and other kinds of place – the gap that the government's 1977 white paper sought to narrow – remains as wide as it was a decade and a half ago. In some respects the gap has widened.
>
> (Willmott and Hutchinson, 1992: 82)

Policy for the Inner Cities was a strategy of targeting public sector programmes on the inner cities, both generally and through specific projects funded by the Urban Programme. It was not this policy which failed; hardly before it had started the new Conservative government elected in 1979 abandoned it. A comprehensive policy based on public expenditure was replaced by a 'patchwork quilt' of different initiatives, with considerable central government control, much less local government control, and a lead role for private developers and business interests (Audit Commission, 1989).

The reason for the abandonment of a comprehensive urban policy was the new Conservative government's faith in market mechanisms and its commitment to reduce the proportion of the nation's income consumed by public expenditure. In fact, the government was not very successful in reducing public expenditure because of the increase in social security payments caused by a large growth in unemployment, and local government expenditure as a proportion of gross domestic product fell only slightly, from 12.3 per cent in 1979/80 to 10.5 per cent in 1988/9. However, public services came under strong financial pressure due to growing needs and cost containment, and some areas such as housing and the Urban Programme were cut back sharply.

A series of indicators revealed continuing and often growing problems in many parts of the country during the 1980s. Examination achievement in schools in the most deprived areas of England fell further behind the rest of the country and showed no improvement in Scotland; homelessness increased in both the country as a whole and the most deprived areas, with particularly sharp increases in London and some other large cities; and the proportion of people dependent on income support rose slightly in deprived areas, despite an overall fall in the country as a whole. Serious problems of unemployment, physical decline and crime emerged in many large peripheral council housing estates to parallel the problems of inner cities.

However, the 1980s also saw improvements. Declines in still births and infant deaths occurred to varying extents in all of Britain's most deprived areas, reflecting a rise in living standards for most poorer people. The relative employment position of some deprived areas also appears to have improved during the 1980s and to be linked to the spending of Urban Development Corporations, City Action Teams and Task Forces (Willmott and Hutchinson, 1992).

There was a small improvement in the real incomes of most poor people during the 1980s, but their relative position deteriorated because benefits have not been increased as much as average earnings. Table 11.1 shows how the value of unemployment benefit declined compared with average earnings over the period 1979 to 1992.

Overall, government figures show that the incomes of the poorest 20 per cent of households in Britain improved only marginally during the 1980s, while the incomes of the richest 20 per cent rose substantially. Table 11.2 shows the average incomes of the richest and poorest 20 per cent of households before housing costs and equivalised for household size and composition.

Table 11.3 shows an analysis of incomes for the poorest 20 per cent broken down by household type. Couples with children and single people who are in the bottom 20 per cent of the income distribution have seen their real

Table 11.1 Unemployment benefit at April 1992 prices (a) and as a percentage of gross average earnings (b)

	Single person		*With dependent spouse*		*Couple with two children under 11*	
	(a)	*(b)*	*(a)*	*(b)*	*(a)*	*(b)*
1979	£42.62	16.2%	£68.99	26.3%	£76.83	29.2%
1980	£41.25	15.3%	£66.72	24.7%	£71.72	26.6%
1981	£40.14	15.1%	£64.94	24.4%	£67.80	25.4%
1982	£41.98	15.6%	£67.92	25.2%	£68.93	25.6%
1983	£43.32	15.7%	£70.07	25.4%	£70.55	25.6%
1984	£43.42	15.2%	£70.20	24.6%	£70.20	24.5%
1985	£44.06	15.3%	£71.27	24.7%	£71.27	24.7%
1986	£43.84	14.6%	£70.88	23.6%	£70.88	23.6%
1987	£42.88	14.0%	£69.33	22.7%	£69.33	22.7%
1988	£42.97	13.3%	£69.47	21.5%	£69.47	21.5%
1989	£42.14	12.9%	£68.12	20.8%	£68.12	20.8%
1990	£41.44	12.6%	£67.01	20.4%	£67.01	20.4%
1991	£43.17	13.0%	£69.82	21.0%	£69.82	21.0%
1992	£43.10	12.8%	£69.70	20.7%	£69.70	20.7%

Note: Comparisons are for November for 1979 to 1985, July for 1986 and April for 1987 to 1992 (the benefits were uprated in these months).
Source: *Hansard* (1992a, 1992b), Parliamentary copyright.

Table 11.2 Average equivalised income at April 1992 prices before housing costs

	1979	*1981*	*1987*	*1988–9*
Bottom 20 per cent	£96	£94	£100	£100
Top 20 per cent	£289	£298	£376	£402

Source: *Hansard* (1992c), Parliamentary copyright.

Table 11.3 Average equivalised income at April 1992 prices for the bottom 20 per cent by household family type

	1979	*1981*	*1987*	*1988–9*
Pensioner couple	£87	£91	£99	£98
Single pensioner	£87	£92	£95	£92
Couple with children	£101	£90	£98	£100
Couple without children	£125	£127	£118	£129
Single with children	£85	£89	£93	£93
Single without children	£109	£107	£107	£107

Source: *Hansard* (1992c), Parliamentary copyright.

incomes decline slightly during the 1980s.

Urban areas have become more socially polarised, reflected in the contrasting affluence and poverty of the leafy suburbs and the inner city estates. This is damaging the social fabric of society as crime, ill-health and concerns about an 'underclass' of people trapped in unemployment mount.

Paralleling these problems is the deteriorating state of the global environment. The world's protective ozone layer is being damaged by chlorofluorocarbons (CFCs). 'Greenhouse gases', principally carbon dioxide produced from burning, may be causing global warming, leading to possible flooding from sea-level rises, desertification from climatic change and mass starvation in areas rendered infertile. The present rate of use of natural resources is unsustainable.

In June 1990, the Dublin Summit of European Community leaders made environmental policy a key element in creating a 'People's Europe'. In the same month, the European Commission published its *Green Paper on the Urban Environment*. This document called for a fundamental review of urban land-use planning and car-based transport policies to reduce urban pollution and traffic congestion, conserve urban heritage, 'green' urban areas, conserve energy and resources, and improve environmental education, training and research.

There are powerful economic, political and environmental influences now shaping urban policy in the UK. The dominant economic influences are deindustrialisation and 'post-Fordist' conditions of high unemployment and a peripheral employment sector of low pay, part-time contracts and casual work. This is sometimes described as the 'one-thirds/two-thirds' society. A large minority are deprived of the comforts enjoyed by the majority, although the dividing line is not strongly defined.

The dominant political influences have been the 'New Right' strategies of economic restructuring and selective rather than universal welfare services to reduce the tax burden. These have seen the introduction of legislation to weaken trade unions, strengthen the hand of management and reduce the benefits delivered by the welfare state. This has affected both the private and public sectors. In the public sector, there has been a strategy of extending competition and privatisation. Private sector ideas have been introduced to reduce costs, and the purchaser–provider split has been accompanied by the rise of a 'new managerialism' which has sought to organise the work of professionals within business plans and performance indicators.

Deindustrialisation has seen an eclipse of class politics based on industrial trade unions. Grand plans for reform based on socialism or social democracy do not exist today; it is difficult to imagine what the contemporary equivalent would be of the Beveridge Report, which launched the British welfare state, the Robbins Report which led to a wave of new university building in the 1960s, or the massive post-war new town building programmes. Grand plans today are often seen as impositions; society is more individualistic and there has been a reaction against uniformity (a criticism levelled at the bureaucratic welfare state, the 1960s universities and the post-war new towns). Individualism has been fuelled by a greater diversity of goods and services, and by New Right politics which celebrate acquisitiveness and competition, and denigrate socialist and collectivist ideas. Society is more fragmented due to a diversity of economic experiences and the rise of 'identity' – expressed both in consumption and in 'identity politics' based on race, culture, religion, gender or sexuality.

In both the political sphere and civil society there has been a revolution in popular attitudes to the environment in recent years. Green consumerism, green economics and green politics have all emerged as strong forces in contemporary society. According to *Eurobarometer*, 85 per cent of Europeans think that the future of the environment is an immediate and urgent problem (*Hansard*, 19 June 1992).

These trends are major challenges for urban policy, but not impossible ones. And there are some trends which are promising. The growth of consumerism in public services has been used as justification for the privatisation strategies of the New Right, but it is also part of a general social trend. The private sector was affected by a growth of consumer movements in the 1970s and has been affected just as strongly by the growth in environmental awareness during the 1980s. Citizens' charters were pioneered by Labour councils. Business planning and performance indicators are essentially variants of the corporate planning introduced under social democratic governments during the 1960s and 1970s. Although many of the trends occurring in society today are inimical to the values of public service, others are positive developments. These include the new emphasis on need assessment, corporate approaches, community development, assessing outcomes, 'customer care', and the greater use made of research. There is no reason why such developments should be used to narrow entitlement to services and benefits rather than to go further in meeting needs and empowering people.

There are also some good prospects for sustainable urban policy. This is largely as a result of European policy. There have been more than 150 European Commission directives and regulations aimed at protecting the environment. Many treaty infringement procedures have been initiated by the Commission against member states: examples include the quality of bathing water, the quality of drinking water, sulphur dioxide in the air, and cutting peat, building a ski lift and depositing sludge in protected areas. Local government can have a central role, and the Association of Metropolitan Authorities has pressed for local authorities to be recognised as local environmental protection agencies (Association of Metropolitan Authorities, 1992).

An area where there are few promising signs, however, is the UK's economic position and the weakness of its manufacturing base. Both urban and regional policies are needed to address these problems, but their scope is limited by both the ideology of successive Conservative governments opposed to state intervention, and local government and regional structures which do not provide effective business support (see pp. 300–3).

URBAN POLICY – MEETING NEEDS?

The concept of need is central to urban policy and is present in all the chapters of this book. The frequency with which the concept is used today reflects the concern with 'concentrating help where it is needed most' at a time of pressure to rein back public expenditure. It would be inappropriate to end a book about urban policy without returning to consider the concept, especially as this leads to the question of how decisions are made about which needs are recognised and the extent to which they are met.

In conventional terms, 'need' is assumed to be an intrinsic, objective and measurable property of individuals (Armstrong, 1982). Urban policy is about 'meeting needs': for example, 'housing needs', 'health needs', the welfare of 'children in need' or disabled people's 'needs'. However, the definition and prioritisation of need can be very contentious, as illustrated by the government announcement in July 1994 that the right of unintentionally homeless families to permanent council housing would be ended by new legislation (*The Times*, 19 July 1994). Instead, local authorities will be required to provide homeless families with temporary housing, mainly private rented accommodation, and then assess their housing needs alongside other applicants waiting for council housing. The plan was a reaction to allegations that pregnant teenagers were jumping council housing queues.

In the early 1970s, Bradshaw (1972) devised a simple classification of social need. This has four categories. *Normative need* is defined by experts, professionals and policy-makers. It is often couched in terms of a deficiency against a desirable standard. *Felt need* is people's own wants, desires or subjective views of their needs. These may or may not become expressed needs. *Expressed need* is felt need which is turned into actual demand for services. *Comparative need* exists when people with similar characteristics to those already in receipt of a service are not themselves receiving that service.

Common to all these categories of need is that they are subjective interpretations based on value judgements. There may not be agreement about what any of these needs actually are in practice. It could be argued that certain needs are basic or 'categorical' because they have to be fulfilled for a person to develop as a human being. These could include health, nutrition and shelter. However, Bradshaw (1992) argues that this still leaves unanswered how to establish the required level of health, nutrition or shelter, and what level of human development is being aimed at. Given that needs are met from limited resources, there have to be priorities about which needs are met first and to what extent.

Central and local government often claim to allocate services according to need, but there is no objective way of doing this. Usually 'needs' in this context are criteria for rationing services, as Bradshaw (1992) describes in his example of the Social Fund. Claimants apply to the Social Fund for a loan or grant if their 'needs' cannot be met out of the scale rates of social security benefits. Bradshaw writes:

> Need is what claimants ask for, if the claimant meets the criteria of the Act and regulations, if they belong to one of the priority groups listed in the guidance to Social Fund officers, if the item they ask for is also considered a priority in that guidance, if the Social Fund officer recognises all this and then only if there is money left in the local office budget.
>
> (Bradshaw, 1992: 6)

This example illustrates how needs are often what the state decides to recognise and provide services to meet. Even in areas where there are explicit commitments in policy to be 'needs-led', very few services are provided on request and what people receive may still largely depend on what providers decide to supply. Miller and Munn-Giddings cite the following example of a 35-year-old woman with cerebral palsy using a wheelchair and caring for her disabled 71-year-old mother, both of whose needs were assessed under the new community care arrangements:

> Both women can manage all their personal care and prefer to do so; they requested help with vacuuming the house, for one hour a week. Instead they have been offered ten hours of personal care and help with housework has been refused. It is likely that in terms of priorities those people with informal carers will be viewed as less in need of services than those that live alone. The growing research on elderly people at risk of 'abuse', however, consistently indicates that those people living with and dependent on their carer 24 hours a day are at highest risk.
>
> (Miller and Munn-Giddings, 1993: 48)

Professional ideology is significant. The widespread use of a medical model to describe disability is an example. This model defines disability in terms of an individual's functioning rather than in terms of disabling features of the environment. The medical model approach is then to try to modify the individual to fit the environment, rather than to modify the environment to fit disabled people. The OPCS national disability survey carried out in 1985 used functionally oriented definitions based on this model. This led Abberley (1992) to make the following criticism of the questions asked by the OPCS interviewers:

> 'Does your health problem/disability affect your work in any way at present?' could be better put as 'Do you have problems at work as a

> result of the physical environment or the attitudes of others?'
>
> 'Does anyone in your household have . . .
> a. Difficulty in walking for a 1/4 of a mile on the level?
> b. Great difficulty in walking up or down steps or stairs?' will tend to systematically underestimate the problems confronted, and often successfully dealt with, by disabled people.
>
> (Quoted in Miller and Munn-Giddings, 1993: 32)

Rennie explains how changing perceptions of need can give rise to very different responses:

> At issue are questions about whether deafness, dwarfism and other disabilities should be regarded primarily as pathologies or as part of the normal spectrum of human variation. Medical opinions evolve over time. Homosexuality was once classified as a mental illness, but psychologists no longer call it one. Alcoholism was formerly a vice; now it is a disease. Accompanying those shifts were changes in attitudes about whether the conditions could – or should – be cured.
>
> (Rennie, 1993: 10)

Aids and equipment should obviously not be withheld and should be further developed and improved. However, they need to be complemented by environmental and social policies, especially as such devices rarely completely compensate for a disability. Rennie (1993: 9) uses the example of deaf people who, 'are entitled to the respect due to any ethnic, cultural or religious minority. The deaf have their own language, customs and history; unfortunately, their eloquence is lost on people who are illiterate in sign language. Because the real problem of the deaf is one of communication . . . it should be solved by a social remedy, not a medical one.'

This perspective moves away from 'head counts' of people in need to consideration of the way society and the environment are organised. Miller and Munn-Giddings comment on Mike Oliver's work in this area:

> [I]n terms of disabled people's needs, rather than individualising the problem, agencies (particularly those associated with local government) should measure their own 'disabling environments'. For example, measures of how much information they supply about services and whether services are accessible or not could create a 'disability index' of local authorities (rather than of individuals labelled as 'disabled') . . . [P]eople are disabled by the environment therefore 'need' is created and served by agencies which would (for example) be better employed ensuring that all new housing and buildings were accessible for wheelchair users rather than counting the number of people who use wheelchairs and planning specialist accommodation for them.
>
> (Miller and Munn-Giddings, 1993: 46)

A similar emphasis on the environment is needed for children. Towns and cities need to be planned and managed so that they are 'child friendly' and free from environmental dangers (Petrie, 1990).

Illich, in Illich *et al.* (1977), wrote that professional 'experts' have a vested interest in *creating* needs which keep them in employment and careers. Quoting him, Armstrong states:

> [T]he professions, in asserting secret knowledge which they alone control, have a special incommunicable authority to determine what people 'need', and claim the power to prescribe services to meet those 'needs', so that 'they not only recommend what is good but actually ordain what is right'.
>
> (Armstrong, 1982: 34)

This type of professionalisation is being challenged by the growth of consumerism in public services, embodied in such strategies as charters, customer care, delegation and quality assurance. But it might be argued that this is essentially a different type of professionalism which recasts professionals as providers of a 'service' to 'customers'.

Many people oppose this sort of consumerism in the public sector because social needs are a question of public debate and political decision. Users of public services often cannot choose to 'shop elsewhere' and it is impractical to provide public services on the basis of a 'shopping model'. In the market place, competition between different producers and retailers means that goods remain on display, unused until they are purchased. It is difficult to imagine hospitals or schools having beds and places unused while customers shop around for the best buy. The fact that public services are not funded to provide this amount of idle capacity means that assuring the quality of services and researching needs is particularly important. It also means that there should be plenty of opportunity for people to have a say in how services are run.

Community development has an important role in developing participation, especially among poor families, minority ethnics, tenants, women, and disabled people. These groups are often the most dependent on public services but the most under-represented in national and local government. The existing structures of local democracy are not sufficiently sensitive to represent the diverse and multi-cultural nature of contemporary urban society. Community development should be a strategy for equal opportunities, supporting more representative, stronger and reliable community organisation in situations where division, poverty or discrimination otherwise impair effective representation. It should complement other strategies to make services more accessible and accountable, principally decentralisation and performance monitoring. For example, community workers can provide support to the residents' side of decentralised area committees, and community groups can be involved in helping to monitor, evaluate and develop council services.

These considerations lead to the type of model for local government suggested by Hambleton (1992). Local democratic control must mean strategic control over services and regulatory functions. The model of control by contract has much to commend it in this respect because it establishes a strategic contract specification and monitoring role whilst allowing a diversity of provision by the public sector, voluntary bodies, community organisations, co-operatives and private firms. Within the policy and resource allocation strategy of the local authority, service providers should be given a high degree of delegated power to run service delivery flexibly and in a way suited to local circumstances. Customer management and community development ensure that services are kept close to their consumers and reflect a diversity of interests.

This pluralism should be reflected in strategic policy itself, which Hambleton (1992) suggests is one of, 'mediation of interest and the management of complexity'. He describes his model as follows:

> [T]he role of the centre is to give strategic direction and encourage the creation of an organisational culture which is committed to the philosophy of the organisation. This model recognises that it is unsound to stipulate from the centre precisely how service delivery is to be specified. It is driven by the belief that the periphery of the organisation – the front line – is closer to the consumer. It is a model of local government which combines strategic direction with decentralised, local management.
>
> (Hambleton, 1992: 11)

Although devolving responsibility to the front line is good management practice, the accountability of public services depends on consumers having some power. Recent Conservative governments have sought to achieve this by allowing choice between, for example, a council landlord and a housing association, or between a school controlled by the local education authority and one which is grant-maintained. Individual consumers, however, have relatively little influence and, in practice, relatively little choice. This is because most public services are still organised collectively; it is considered a waste of resources to provide a wide range of individual choice like a supermarket. An important means by which consumers of local government services exercise influence is by electing councillors to whom the services are accountable. Removing this local democratic control puts ordinary people into a weaker position, with no elected councillors to contact with problems or complaints.

The role of councillors as consumer advocates is an extremely important one. Local authorities already play an important part in consumer protection, encompassing trading standards, environmental health, land use planning and building controls. Councillors spend much of their time dealing with problems and issues raised by members of the public relating to local services. This role is one which may be undermined if the contract culture goes too far. There is a danger that once a contract has been let there will be little opportunity to influence service delivery until the contract comes up for renewal. This could well make a service less responsive to the needs of the public than one delivered directly by the council without a contract.

Contracts for public services have begun to address this issue by being less specific about *outputs* and placing more emphasis on *outcomes*. In community care, for example, providers might not carry out tasks identified in the service specification because the user asks them to do something different. An organisation might be contracted to bathe a person living in their own home. But a user might say that she or he would rather the care worker take them out to the shops for an hour. This outcome might be as satisfactory as bathing. It would be very inflexible to insist that the care worker's task must always be to provide a bathing service. The prime concern is an outcome for the user which enables them to maintain their quality of life remaining in their own home, without admission to residential care.

THE ISSUE OF LOCAL DEMOCRATIC CONTROL

Education policy illustrates very sharply the problem of 'getting the balance right' between different levels of decision-making: in this case between central government, local government and schools. During the 1960s and 1970s the education departments of local authorities dominated school provision, managing schools and making large investments in professional and curriculum development, buildings, and teams of well-paid inspectors and advisers. Many local education authorities targeted substantial resources on schools in areas of deprivation.

Education legislation during the 1980s, continued by the 1993 Education Act, shifted power towards the schools in a system where 'market forces' through parental choice and/or selection by schools now determines whether schools grow or decline. Schools have control over their own budgets, allocated to them by the local authority using a formula based predominantly on the number and age of pupils (see Chapter 4). Responsibility for inspecting schools has been largely transferred from local authorities to a new quango, the Office for Standards in Education (OFSTED).

The key responsibilities that remain with local authorities are largely in the welfare area: ensuring that all children receive education, securing special education, assessing and statementing children with special education needs, and certain pupil specific and support services. The major area of strategic planning only remains a local government function if less than 75 per cent of pupils are in schools that have opted out of local authority control. After this point, responsibility for providing sufficient school places passes to the government's Funding Agency for Schools. This also includes identifying those unpopular or otherwise unviable schools where there is a need for rationalisation.

Many schools have welcomed the greater freedom to manage their own affairs brought about by the devolution of budgets. However, the loss of local democratic accountability is substantial and in many areas control has passed from local to central government, from national testing to requirements about what information is reported to parents. New bureaucracy has been introduced to undertake functions previously carried out by local authorities, principally OFSTED and the Funding Agency for Schools. Morris (1993) has called these new bodies the 'magistracies and commissariats' of education. Relegation of the local authority role has had the effects of seriously undermining accountability to local electorates, the capacity to plan housing, social services, education and other services together, and the ability to provide a 'seamless' service from pre-school to adult education.

Whether centralised or decentralised, purchasing services or providing them, local government is founded on the principle that elections can change things. A national survey by John and Bloch (1991) of public attitudes to local government found that 68 per cent of respondents thought that voting in local elections could change policies and services. This was, however, a decline from 77 per cent when the same question was asked in a 1965 survey. Where a particular party has held power for decades, such as in the case of the Labour Party in many metropolitan areas, local electors are less likely to believe that their vote can bring about change.

Despite the strategy of recent Conservative governments to reduce the powers and influence of local authorities, local government retains support among the British public. John and Bloch (1991) found that 77 per cent of respondents thought that their local authority ran services 'very well' or 'fairly well'. A substantial majority in this survey was in favour of councils providing services directly rather than bringing in outside contractors, including having the lead role in providing social housing for rent. Similarly, public opinion seems to be against central government measures to restrict the freedom local councils have to make their own decisions, including setting local taxes and deciding on expenditure levels.

However, there is evidence that public opinion has become more divided between those who favour greater central government control and those (still the largest proportion) who favour less central government control. In addition, whilst most people still believe in local elections, more are inclined to think that local councillors forget their election promises or that local elections do not affect the way people live. Thus, the general support that exists for local government in Britain must be qualified by this evidence of some trends towards polarisation and cynicism about local democracy.

The decline in support for local democracy makes the job of a local councillor more unrewarding, leading to less people being interested in becoming an elected representative, further undermining the vitality of local government. Local councillors spend around 60 to 100 hours per month on council business, for which they receive small allowances. Reconciling these demands with paid employment or family responsibilities can prove extremely difficult. At each local election, about a third of councillors drop out from the council (Bloch, 1992). Only 25 per cent are defeated at the polls; most of the remainder drop out for personal reasons after only one or two terms. The most important single personal reason is the competing demands of work and family.

The new model for local government advocated by Hambleton (1992) leads him to argue that local government needs more councillors. However the Local Government Commission established to review the structure of English local government has tended to reduce the number of councillors by recommending replacement of the two-tier structure of district and county councils with a single tier of all-purpose authorities. For example, recommendations published in January 1994 for replacing the two-tier system of two county and seventeen district councils with seven all-purpose authorities in Yorkshire and Humberside have this effect. The commission argued that, 'by reducing the number of councillors, it will force the introduction of an improved support network for councillors, enabling effective representation across the review areas' (March, 1994: 15). The commission has proposed that the ratio of residents to local councillors should be 4,000:1, which compares with a current ratio in England and Wales of 1,800:1, already much higher than in other European countries where it is less than 500:1 (Jones and Stewart, 1994).

Government ministers are on record as saying there are too many people involved in local government decision-making. The committee process of decision-making in local government involves all councillors in both policy and administrative details, even though in practice it is usually a fairly small number of councillors who have leadership roles. Michael Heseltine, when Secretary of State for the Environment, made the following criticism in 1991:

> The problems that local authorities face are compounded by the cumbersome internal arrangements for the management of councils. The committee system – which dates back to the Municipal Corporations Act of 1835 – requires that all decisions are taken collectively by large numbers of councillors, in full council or committees. Too many councillors spend too much time achieving too little.
>
> (*Hansard*, 1991, col. 402)

British local government is unusual in not having a clear distinction between a majority of councillors who perform in representative, non-leader roles in the council, and a minority who comprise the 'executive' responsible for political direction and leadership and co-ordinating the various tasks of the local authority. This model has much to commend it in terms of providing political direction for the authority, making clear where accountability for decisions lies, providing a more efficient and co-ordinated decision-making process, and providing a forum for policy options to be discussed with official advisers. An executive model of decision-making has already been introduced informally by some authorities, but formal adoption of a separate executive committee actually empowered to take decisions on behalf of the council would require legislative change. In 1993, a report by the Department of the Environment's Joint Working Party on Internal Management considered various models for executive decision-making, such as a single party executive committee or a cabinet system (Department of the Environment, 1993b). It stressed that such models should be balanced by strengthening the representative and scrutinising roles of the majority of local councillors who would not be part of such a centralised arrangement, otherwise local government as 'community government' would be undermined. Ways of doing this include decentralisation and local management of services, and the involvement of 'back-bencher' councillors in panels or 'select committees' which scrutinise the work of departments.

Many local authorities have sought to increase their local community government role. Instead of centralising management, service delivery has been decentralised together with decision-making power to local area or neighbourhood committees (Lowndes, 1991). In some authorities, these local committees constitute full committees of the council. In other authorities, they are advisory neighbourhood forums with community groups and service users represented.

The general aim of decentralisation is to bring the council and its services closer to people. During the 1986–94 period of Liberal Democrat control in the London Borough of Tower Hamlets, standing neighbourhood committees *replaced* departmental central committees as the main decision-making forums for policy and budgets. One consequence was that *more* councillors were involved in decision-making, and some of the committees were actually controlled by the opposition party.

Combined with the one-stop shop principle, Tower Hamlets' decentralisation to neighbourhoods appears to have seen residents' satisfaction with the council increase (Brooke, 1994). The creation of an internal market and contracting out of services, together with a flatter organisational structure in the neighbourhoods, kept costs down and enabled the decentralised structure to run at no extra expense. However, there can be problems with this degree of decentralisation. The planned and equitable distribution of resources becomes more difficult because control is so localised. In Tower Hamlets, the neighbourhoods had different lettings policies and targets for housing; they could also choose to opt out of borough-wide schemes such as transfers. Tower Hamlets also had to make special appointments to carry out limited central functions for its services and to market the borough as a whole. The new Labour administration, which succeeded the Liberal Democrats following the May 1994 elections, disbanded the seven neighbourhood committees. Although four community committees were formed for service delivery and consultation, decision-making was returned to a series of central committees and a single chief executive was re-established.

Decentralisation and approaches such as quality assurance and performance indicators have been pursued as strategies to improve local government services. However, changes in how public services are managed can only partly address urban problems. These include the ageing of Britain's population and increasing numbers of people who need social care, environmental damage and the need for policies for conservation and sustainable development, and the state of urban infrastructure – roads, public transport systems, schools and housing – which is in urgent need of maintenance and improvement. Jeremy Beecham, leader of Newcastle City Council, recently commented:

> What we are trying to do here is to mitigate the social damage, the chronically high levels of unemployment, spinning off into poverty, the crime, the alienation ... meanwhile the Government in London remains caught in a vicious downward spiral that started in 1979, restricting us in a way that is profoundly damaging for government and for local democracy and perhaps for democracy altogether. It is working against us when it should be working with us ... We are trying hard to keep the lid on, to keep the Elastoplast applied ...
>
> (Quoted in McPhee, 1993: 13)

The response to these growing pressures has not been to expand public expenditure proportionately but to strengthen the targeting of public expenditure and reduce the costs of public services. Increasingly, public services are means-tested or targeted on those most in need using techniques such as individual assessment, priority areas or community development. For example, it is not possible any more to provide publicly funded social services to all older people who could benefit from care. Today, social services departments have to target the public funding of care to people most 'at risk' and those with the highest levels of dependency. Many of these services are also means-tested to assess whether the user should pay a financial contribution towards their cost.

A consequence of greater targeting is that publicly funded provision becomes a 'fire service' responding to actual or potential crises in people's lives. However, failing to provide adequately for the welfare of those with medium risk or medium dependency will not only impair the quality of life such people could have but will also leave them vulnerable to becoming high risk because of a lack of support earlier on (Means, 1992).

The undermining of local government's ability to respond to local social and economic conditions reflects a wider breaking down of consensus about the role of local government. In particular, is local government to continue along the road of becoming little more than an administrator of some public services, or should local authorities be recognised as community government by local citizens participating in the local democratic control of major public services (Hambleton and Hoggett, 1990)? The conflict of political opinion about local government's role has contributed to an increasing politicisation. In 1986, the Widdicombe Committee reported that the last twenty years had seen the number of councils not controlled by political parties drop from 50 per cent to only 14 per cent (Widdicombe Report, 1986). The Conservative Party's policies brought about sharp differences between the main political parties over compulsory competitive tendering, council tenants' 'right to buy', and particularly levels of spending on services and control over local taxation. These differences are sharpest in the large urban areas where support for public services and opposition to market economics is strongest.

Consumerism is a valid and welcome influence in the public sector, but it fits well with an agenda for local government as little more than the administrator of contracts for some local services. By contrast, the community government role is one where a local authority leads in 'the process of taking the overview of the needs and demands of its area and people and then seeking, by both direct provision and by working through other bodies, to meet the needs and satisfy the demands' (Alexander, 1991: 74). Despite these tensions and the erosion of local government functions and autonomy in recent years, the case for strong and stable local government is increasingly being reasserted. Martin Easteal (1993), chief executive of the Local Government Commission for England, has identified four reasons why this is necessary:

1 Local democracy is good for democracy generally, involving local citizens in decisions about their communities and enabling them to balance conflicting interests locally without reference to central authority.
2 Local policy-making creates opportunities for testing and evaluating alternative ways of providing services, without committing the whole country to a particular course of, possibly erroneous, action.
3 Returning functions transferred to other agencies such as urban development corporations back to local authorities should save money.
4 No other European country is reducing the powers of local authorities, and many are increasing them.

The concept of subsidiarity adopted by the European Union has been used by local government in Britain to press its case for increased powers. Norton defines the concept as follows:

> The doctrine of subsidiarity ... lays down that the responsibility for carrying out tasks should be held at the lowest level of government competent to undertake them, and that where necessary higher

> authorities should give support to enable them to fulfil the responsibilities that are appropriately theirs under this doctrine.
>
> (Norton, 1991: 27)

This principle is behind the increase in powers being given to local government in many other European countries. The British government has laid particular emphasis on the principle to protect its powers in relation to Brussels, but domestically has pursued the opposite in removing responsibilities from British local government.

Constitutionally, the erosion of local government is alarming because local democracy is a counterweight to the comprehensive powers of central government. In the UK, central government powers are not strongly vetted by parliament. In addition, the centralisation of power means that responsibility for more and more services falls to government ministers, including an estimated 40,000 appointments to quangos (Beavis and Nicholson, 1993). Ministerial workloads become distorted, local accountability is lost and central bureaucracy grows. This is despite the absence of any evidence that local councils provide services less effectively than central government or appointed agencies. Indeed, the latter add to the fragmentation of urban policy.

A further concern about the loss of local government powers in Britain is the resulting loss of effective city leadership, increasingly recognised as an essential ingredient behind the dynamism of many European and North American cities. As the Association of Metropolitan Authorities comments:

> This is a deeply uncomfortable agenda for local authorities trying desperately to provide good government, maintain service provision and develop the economic, social and physical attractiveness of their cities against competition not only from within Britain but from Europe.
>
> (Association of Metropolitan Authorities, 1992: 2)

However, which level of government should be responsible for which functions, and with what degree of policy-making autonomy, are difficult issues. Policy formulation needs to be sensitive to local variations, apparently a clear justification for local government. But after the Second World War, a Labour government removed from local authorities their last residual income maintenance powers and control over the municipal hospitals as part of its strategy for building a national welfare state, attempting to end local variations by introducing national standards of provision. In the 1990s, there appears to be more emphasis on localities and their different economic, political and cultural identities which some commentators argue local government should reflect (Gyford, 1991).

A common criticism of the 1974 reorganisation of local government in England is that it created new county councils, notably Avon, Cleveland,

Hereford & Worcester and Humberside, which are artificial administrative units that do not reflect local identities. A variant of this approach is that local government should also be organised so that there is clear accountability to people living in a particular locality. In 1986 the government abolished the Greater London Council and the six 'upper tier' metropolitan county councils in England citing as one of its reasons the objective of making local government in the big cities more clearly accountable to local people. As a result, local government in the major English conurbations consists of a single tier of all-purpose metropolitan district councils and London boroughs.

In most other parts of Britain, 'unitary' all-purpose local authorities are likely to replace the present two-tier structure of district councils and county councils (regional councils in Scotland). The reviews of local government structure begun in 1992 by the Local Government Commission in England and the Scottish and Welsh Offices are due to be finished at the end of 1994. These are likely to result in breaking up county councils or merging smaller district councils, although there is no national blueprint. For Lincolnshire, for example, the Local Government Commission recommended no change to the existing two tier structure. It has been claimed that the simplification of local government into one tier will help reduce public confusion about 'who is responsible for what'. However, there is a danger that the creation of new non-elected joint authorities to oversee services that have to be run over areas wider than those of smaller unitary authorities, such as highways and waste disposal, will blur lines of responsibility and accountability rather than clarify them in the public's mind.

REGIONAL GOVERNMENT

Today, all EU member states of a similar size to the UK have either a federal structure or a constitutionally prescribed system of regional government. Devolution from central government to regional authorities has been a recent trend in most European countries, with the notable exception of the UK. In 1986, the first direct elections were held to twenty-two new regional governments in France. These took over responsibility from central government for secondary schools and vocational training. Central government's prefects (its representatives in the provinces) had their power to veto local decisions removed. In Spain, a strategy of rolling devolution to seventeen regions or 'autonomous communities' is transferring power from the centre to the regions. A similar process is starting in Portugal. In Italy, new tax-raising powers have been given to local authorities and regional devolution is high up the political agenda. West Germany's highly successful system of regional government was established by Britain and her allies following the Second World War to check any resurgence of German nationalism. The system has a financial balancing mechanism which favours the less prosperous regions in public spending allocations. It was successfully adopted in East Germany following unification. Norway, Sweden, Denmark and Finland are increasing local autonomy by relaxing central controls on local authorities.

A recent study concluded that decentralised states like Germany are more likely to have successful economic policies:

> Indeed, there seems to be something about population units of around three to eight million, the size of many regions of the larger European states, or smaller European nation-states, that makes policy-making between public authorities, business organisations and interest groups particularly useful and flexible.
>
> (Tomaney and Turnbull, 1992: 13–14)

Germany's federal system of government involves extensive powers for regional governments, including legislative powers. In industrial policy, these regional governments actively intervene to manage industrial restructuring and modernisation in order to protect employment levels.

In England, the centralisation of decision-making, reduction in local government powers and functions, transfer of other powers and functions to local agencies appointed by central government, and narrow financial autonomy of local authorities mean that the United Kingdom is very close to having in effect only one tier of representative democracy.

In Scotland and Wales, the picture is complicated by a trend towards *administrative devolution* by transferring more powers from London to the Scottish and Welsh Offices. The Welsh Office, for example, was a very small body when it was established in 1964, but today is responsible for over 70 per cent of public expenditure in Wales (Osmond, 1993). A large proportion of this, however, is spent by appointed bodies at the cost of local government (see Chapter 3). In England, there is an important tier of regional administration but it is not accountable to regional electorates. This includes regional offices of the NHS and integrated regional offices of central government.

The absence of regional government and the erosion of local government in the UK have disadvantaged regions outside the south east of England in two ways. *Politically*, there is less democratic accountability. *Economically*, there is dependency on sources of investment and finance which are based in London or overseas. Regionalisation is not just about devolving powers to a regional tier of government to oversee regional policy. It is also about having power decentralised politically in regional governments and economically in regional development banks and industrial clusters based in the region.

Regional planning was abandoned at the end of the 1970s and regional policy has been massively scaled down compared with its size during the 1960s and 1970s. Instead, state intervention has been focused on the sub-regional activities of urban development corporations and their property-led strategies for renewing old industrial areas. UDCs have subsidised major redevelopment projects which have often created areas of offices, housing and recreational facilities where there was once manufacturing industry. This free market approach contrasts strikingly with the German example of Hamburg's Development Agency which may refuse planning permission to industries which are not in line with its strategy of generating a high proportion of skilled jobs in an area of relatively high unemployment (Coopers and Lybrand/Business in the Community, 1992). These considerations are far from evident among government-appointed industrial development agencies in Britain, which have sought to attract new factories owned by multinational companies regardless of such questions as the skill content of the jobs. These strategies of market-led regeneration and attraction of overseas multinationals have essentially been at the expense of supporting the modernisation and expansion of the country's traditional indigenous manufacturing sector (Tomaney, 1993).

The limited powers which exist in the UK to make key infrastructure improvements at local and regional level have recently been criticised as an obstacle to economic development (Coopers and Lybrand/Business in the Community, 1992). Similar problems are caused by the lack of a regional banking system and of regional development bodies that can undertake detailed sector and geographic analyses and plan for economic development. Cooke (1993) identifies how such regional structures create 'information rich' regions, where government and industry can be in close contact, problems identified early and collaborative solutions worked out in partnerships between government, small firms and large companies.

Regional government has also been a means by which regions, and sometimes nationalities, have sought to assert their cultural identities. This is evident in contemporary campaigns for regional assemblies in Scotland, Wales and the North of England. However, regionalism is very different from the militant nationalism that has seen a bloody resurgence in Eastern Europe with the collapse of communism, or the sectarianism of Northern Ireland. It is strongly influenced by the idea of a 'Europe of the Regions'. Modern regionalism is described by O'Neill as having the following origins and purposes:

> One of the great achievements of what we used to call the Common Market was in bringing together countries whose history is of military conflict with each other: notably France and Germany. An enlarged Europe, embracing the principle of subsidiarity, not as some magic formula which will cause nationalism to disappear, but as a system of devolved decision making – of democratic accountability – secured by the rule of law, could offer an alternative vision to one of incessant ethnic conflict. Such a Europe of the regions could conceivably transform militant nationalism into a potent force for cultural and

> intellectual diversity. The Europe of the 90s and beyond could also result in a European Union which would have the political will, matched by the economic resources, to help the Third World liberate itself from the evil of underdevelopment.
>
> (O'Neill, 1992: 26)

Of all the main political parties, the Liberal Democrats have the strongest commitment to introducing regional government into England and home rule for Scotland and Wales. This policy complements a local government policy of significant devolution of power to unitary local authorities, including education, health and planning. The Labour Party's policy is separate assemblies for Wales and Scotland and, at a later stage, regional assemblies for England. Conservative government policy is a unitary state with a degree of administrative devolution and, in most parts of the country, a single tier of local government.

Regional government would not be a devolution of power if it acquired responsibilities from local rather than central government. Obvious candidates are the functions already carried out by regional offices of central government, including the strategic planning of housing investment, the NHS, industrial development, transport, and post-16 education and training. A strong role in economic development would necessitate regional banks to assist with implementing industrial strategy. The German regional banking system has worked well in this respect (Tyne Wear 2000, 1987)

SUSTAINABILITY

Urban policy should be about meeting needs democratically. This book has argued for a corporate strategy across all service areas, applying approaches based on needs assessment, research, strategic resource allocation, enabling, quality assurance, community development, health for all, sustainability, equal opportunities in education and economic development. A key role for corporate business plans in local authorities is to achieve integration between individual service plans, linking services into common objectives by assessing and then monitoring their contribution to these objectives.

What these objectives are is a question of political choice. The approach, however, should be one of establishing what desired outcomes are, rather than just stating inputs and outputs for services areas. Thus, if a healthy city is a desired outcome, what are the contributions of housing services, economic development services, the engineering department, the education department, and so on, to this objective? How will this be measured? How will the consequences of actions aimed at achieving this objective be evaluated?

This approach is still fairly uncommon in the public sector because of the day-to-day concerns with delivering services in line with statutory requirements and within budget. But the growing evidence of damage caused by urbanisation and industrialisation to the capacity of the environment to support human activities is making such an approach essential. It is now widely seen as unacceptable that the present generation should meet its needs by compromising the ability of future generations to meet their needs. Sustainable development is a core set of objectives for urban policy and presents the strongest possible justification for strong corporate policy. It is an outcome to which all public services and regulatory functions should be contributing, and an area where the need for a business planning approach of targets and measurable indicators is most pressing.

British public services have been subjected since the late 1970s to stringent cost containment and privatisation policies in order to prioritise economic growth. Sustainable development has to replace unsustainable economic growth as a priority, and this could see a very different attitude develop towards state intervention and social policy. Concern for future generations – or 'intergenerational equity' – lacks coherence if the same concern is not extended to a fair distribution of resources among people already living. Poverty often forces people into unsustainable behaviour. The UK local government associations have suggested that a new agenda has emerged (Association of County Councils *et al.*, 1993). They explain this as follows:

> Sustainable *development* is not the same thing as *economic growth*. The concept of 'quality of life' recognises that standards of living cannot be measured by purely economic indicators or delivered by simple quantitative growth of income. Factors such as the quality of the environment and the overall *health* of the population – which are in themselves closely linked – must also be considered. Quality of life is difficult to measure, and its components will not always be agreed. But this demands further work on attempting to measure it and to gain consensus, rather than claiming that income growth is therefore the sole or best measure of development.
>
> (Association of County Council's *et al.*, 1993: 6)

Evidence from quality of life studies suggests that conventional indicators of need may be a poor basis for formulating policy. Health, and family and neighbourhood stability appear to be of greatest importance to most people (Ley, 1983). Although freedom from poverty and deprivation is essential to realise these aims, increases in monetary incomes on the basis of unsustainable economic growth is clearly not desirable. In addition, as was noted in Chapter 4, even the present level of public services and social security benefits cannot be sustained without further increases in taxation. What is necessary is a fairer distribution of resources in a sustainable world. The alternative is social and environmental instability arising from social division and environmental damage. Health inequalities are a powerful indicator of how far we have to go to achieve a less stressful and fairer society.

There is also a need for greater understanding of complex natural and social systems so that more certainty can be established about the consequences of different actions. Research and development is thus essential to drive policy in the right direction, even if exact answers can never be found.

These concerns with quality of life, equity and understanding are not possible alternatives to market economics but essential responses to environmental threats. They involve assessing 'needs', but do not have objective or scientific answers: 'They can only be settled through open and democratic processes of consultation, collaboration and consensus building' (Association of County Councils *et al.*, 1993: 10).

APPENDIX 1
Leicester 'Environment City' Agenda 2020

The strategic objectives of Agenda 2020 are presented in the eight Specialist Working Group classifications:

THE BUILT ENVIRONMENT

Aim

To create a safe, attractive, pleasant, vibrant and sustainable city through good design and planning of the urban fabric.

Themes

Repopulating the city centre

Encouraging more people to live in the city centre in order to efficiently use vacant office/shop space, create environmentally beneficial travel, minimising spatial distribution and to create a more vibrant city centre social environment.

'Green' development

Conserving and re-using old buildings. Ensuring new buildings use the least environmentally-damaging materials, using energy efficiently, are attractive and are designed for durability. Integrating design with nature.

A safe city

Use the principles of 'designing out crime' wherever possible.

ENERGY

Aim

To minimise demand for energy, promote energy conservation throughout the City and produce as much energy as possible using renewable sources.

Themes

The industrial sector

Use energy audits and all appropriate means to ensure businesses and organisations use the minimum possible energy in their operations and premises.

The domestic sector

Apply energy ratings to all homes, introduce state-of-the-art insulation, heating systems and energy-efficient appliances in place in all homes.

Energy supply

Use renewable resources and passive solar systems wherever possible, and make-up the shortfall using CHP, energy from waste, and any new, clean CO_2 minimising technology that emerges.

TRANSPORT

Aim

To minimise the need for travel and reduce dependence on the car. Where travel does take place, prioritise walking and cycling for short journeys and the provision of quality public transport for longer trips.

Themes

Restructuring the City so as to minimise the need for travel. Provide an extensive network of safe cycle and pedestrian routes. Provide extensive, high quality and affordable public transport systems.

Restricting the car through traffic calming, reduction in parking provision, avoiding road building, and financial disincentives to car use. This will also create safer and more pleasant pedestrian environments.

THE NATURAL ENVIRONMENT

Aim

To preserve and enhance nature conservation and open space in the City and to maximise opportunities for people to enjoy green areas.

Themes

Expand the (already) extensive network of well landscaped 'green' open spaces so green space is easily accessible to all. Continue to carry out detailed ecological surveys and use these as a basis for protecting and enhancing the City's natural habitats and the species they support. Encourage ecologically sound management practices for private and public space. Carry out extensive tree planting, and establish many community woodlands.

Favour native species and eliminate use of peat and undesirable chemicals.

FOOD AND AGRICULTURE

Aim

To minimise the environmental impact of food production and processing; and promote a diet and system of food production which is sustainable and healthy.

Themes

Promote food production in the City (growing at home, allotments, etc.) and the consumption of locally produced food. Promoting sustainable food production – the use of organic growing techniques, permaculture and the adoption of environmentally sound agricultural practices generally. Promote a diet which is healthy and based on sustainably produced food.

THE SOCIAL ENVIRONMENT

Aim

To create a participative society that is environmentally aware and rich in culture. To enhance the City's quality of life and minimise the burden it places on the global environment.

Themes

Developing thorough and imaginative environmental education in schools and colleges, and promote education that encourages self-belief and citizenship.

Raising environmental awareness of all citizens, showing how environment relates to their life (e.g. sustainable health) and inspiring them to act.

Encouraging participation through individual and neighbourhood action, liaison with and support for community groups, and by further creating open and accessible local government.

Promoting the values of quality of life and a buoyant community, and working with national and European government to implement change.

WASTE AND POLLUTION

Aim

To eliminate waste and pollution by reducing demand, re-using and recycling waste, and using best available technology wherever possible.

Themes

Reduction of waste generated by designing for longevity, reducing packaging, and promoting the avoidance of unnecessary consumption.

Promoting re-use wherever possible, and (if not) the recycling of waste. It will be made easier and more attractive to recycle than to throw waste away.

Developing large-scale composting schemes across the city, and using unrecyclable waste for energy production.

Minimising industrial pollution through the fitting of best technology, reducing demand for goods whose production causes pollution, and the imposition of pollution taxes.

APPENDIX 2
Leicester 'Environment City': examples of projects underway

The following selection of projects has been classified according to the eight specialist working groups of Environment City.

NATURAL ENVIRONMENT

Twenty-three projects are presently underway including:

The City Ecology Strategy which has outlined a City-wide network of greenways and natural habitats.

'The Community Woodlands Scheme' has been established, as a ten year programme to double Leicester's woodland cover for nature, amenity, recreation, play and education.

A Garden Survey of seven sample areas of over 4,000 private gardens has been completed to establish an audit base on garden use.

The 'Woody Tree Campaign' has been launched by local media to secure funding for tree planting in Leicester's community woodlands. £75,000 will be raised by the people of Leicester. A progressive ban on the use of peat has been implemented.

BUILT ENVIRONMENT

Sixteen projects are underway including:

The historic core of Leicester is now promoted, restored and enhanced through joint agency schemes such as the Castle Park Management Plan, Town Trail, Heritage Park, the City Centre Action Programme.

Continued resistance to the development of out of town shopping and a positive programme of city centre investment leading to a major reinvestment in the central retail area. A £20,000 feasibility study into 'Advanced Housing Design and Layout' has begun. This will ensure that Leicester retains its lead in the housing field with environmentally friendly design. A 'Green Development Guide' is underway to promote sound environmental techniques in material choice, construction, site layout, management, design and landscape.

Leicester City Council and Leicestershire County Council have banned the use of tropical hardwoods (other than those from approved and specified plantations) for new public sector schemes and grant aid proposals.

ENERGY

Thirteen projects including:

The environmentally friendly 'Eco House' which attracted over 12,000 visitors in 1991.

Combined heat and power units have been installed in two city leisure centres. Coal-fired boilers are being replaced with gas.

Leicester is investigating energy supply and demand and is developing a computer model on energy flow for cities as part of the Energy Cities programme with Barcelona.

Leicester has successfully been nominated as one of Europe's 'Energy Cities' – a network of energy and environmentally conscious cities.

Energy is to be generated from waste at one of the city's refuse disposal sites. This will save methane from entering the atmosphere and provide energy for 2,000 homes with an equivalent output of 3 MW.

TRANSPORT

Twelve projects including:

A new 37 km pedestrian and cyclist 'Green Ringway' which will circumvent the city, linking parks and open spaces. Britain's first 'Woonerf' – a traffic free zone – was pioneered in Leicester. Costing £200,000, 80 properties benefited from the creation of a pleasant, attractive and safe street environment.

A city-wide programme of traffic calming is now underway. A new passenger rail service is being built to link Leicester, Loughborough and Derby.

A 'Greener Driving Campaign' for motorists begins in 1992. Britain's first Green Motor Show will be held in Leicester during 1993.

Leicestershire County Council has produced a new 'Transport Choice Strategy' to outline quality, efficient use of space and environment friendly transport options.

SOCIAL ENVIRONMENT

Eleven major projects are underway including:

Involvement with Leicester's multiple faith groups has brought the International WWF Faith and the Urban Environment Conference to Leicester in September 1992.

Leicester now hosts one of the largest national environmental festivals called 'Think Green'. An Environment Arts Festival and a No Car Day were just two of 1991's activities.

The Environment City 'Community Partnership', launched in September 1991, provides a network for the distribution of grant aid, information and community projects. A data base of other available grant schemes and voluntary bodies in Leicester is currently being prepared.

Leicestershire hosts a unique scheme – 'School Nature Areas in Leicestershire' involving children, teachers and parents in environmental education. A private sector sponsor awards a prize for the school making the best curricular use of a school nature area. Children are also getting involved in environmental design projects (e.g. new recycling facilities).

New education packs on the 'Natural Curriculum' for teachers and children are being prepared to support environmental education.

ECONOMY AND WORK

Eight projects are underway including:

Environment City's unique 'Business Sector Network Launch' which attracted over 100 business delegates and introduced them to good environmental management.

An unprecedented number of Leicestershire companies are involved in a programme of company environmental audits managed by Environment City.

Local business enterprise centres are examining current practice, providing advice on environmental management and lobbying to gain support for Environment City projects.

The declaration of Industrial and Commercial Improvement Areas in Leicester has enabled both environmental and service improvements to 167 premises at a cost of £923,000.

Leicester Ecology Trust's 'Greenworks' project is providing grant support to inner city businesses for the improvement of their premises through ecological planting.

FOOD AND AGRICULTURE

Eight projects including:

A directory of 'Where to buy Peat Alternatives' has been prepared following Leicester City Council's ban on the use of peat in 1989.

A directory of organic food outlets is underway and a campaign to encourage home grown food from both gardens and allotments is being formulated.

The management and promotion of one of Leicester's largest parks is now being developed using Shire Horse technology.

Comprehensive grazing regimes have been agreed for the City's grazing meadows.

Many of the City's parks and open spaces are now cut and baled once a year. The product is sold for bedding or feed to local outlets.

August 1992

BIBLIOGRAPHY

Abberley, P. (1992) 'Counting us out: a discussion of the OPCS disability surveys', *Disability, Handicap and Society*, 7, 2: 139–56.

ACC/ADC/AMA (Association of County Councils/Association of District Councils/Association of Metropolitan Authorities) (1993) *Health of the Nation – Guidance*, 26 July (AMA ref. PW/es/SS Circ 49/1993).

Alexander, A. (1991) 'Managing fragmentation – democracy, accountability and the future of local government', *Local Government Studies*, 17, 6: 3–76.

Allen, T. (1991) 'Unbundling Berkshire', *Local Government Chronicle*, 13 September: 18.

Armstrong, P. F. (1982) 'The myth of meeting needs in adult education and community development', *Critical Social Policy*, 2, 2: 24–37.

Association of County Councils, Association of District Councils, Association of Metropolitan Authorities and the Local Government Management Board (1990) *Environmental Practice in Local Government*, 1st edition, Luton: Local Government Management Board

Association of County Councils, Association of District Councils, Association of Metropolitan Authorities and the Local Government Management Board (1992) *Environmental Practice in Local Government*, 2nd edition, Luton: Local Government Management Board.

Association of County Councils, Association of District Councils, Association of Metropolitan Authorities, Association of Local Authorities of Northern Ireland, Convention of Scottish Local Authorities, Local Government International Bureau and the Local Government Management Board (1993) *The UK's Report to the UN Commission on Sustainable Development: An initial submission by UK local government*, Luton: Local Government Management Board.

Association of Metropolitan Authorities (1992) *Policy and Financial Prospects 1993/4*, London: AMA.

Association of Metropolitan Authorities (1993a) 'Revenue Support Grant and the Unified Budget', Report tabled at AMA Policy Committee, 9 December.

Association of Metropolitan Authorities (1993b) *Local Authorities and Community Development: A Strategic Opportunity for the 1990s*, London: Association of Metropolitan Authorities.

Association of Metropolitan Authorities (1993c) 'DFE Review of the AEN Index', A paper by the Association of Metropolitan Authorities for the Local Government Finance Settlement Working Group, Standard Spending Assessment Sub-Group.

Association of Metropolitan Authorities (1994) '1994/95 RSG Consultation: Changes in SSA', Letter from Association of Metropolitan Authorities to SSA contacts in

metropolitan authorities, 20 January.
Audit Commission (1988) *The Competitive Council*, London: Audit Commission.
Audit Commission (1989) *Urban Regeneration and Economic Development*, London: Audit Commission.
Audit Commission (1991a) *Healthy Housing: The Role of Environmental Health Services*, London: HMSO.
Audit Commission (1991b) *A Rough Guide to Europe: Local Authorities and the EC*, London: Audit Commission.
Audit Commission (1992) *Citizen's Charter Performance Indicators*, London: Audit Commission.
Audit Commission (1993a) *Passing the Bucks: The Impact of Standard Spending Assessments on Economy, Efficiency and Effectiveness*, London: HMSO.
Audit Commission (1993b) *Staying on Course: The second year of the Citizen's Charter indicators*, London: Audit Commission.
Avery, G. (1991) 'Editorial', *HFA 2000 News*, 14: 1–5.
Ayre, D. (1979) 'Instinctive Socialism', in Durham Strong Words Collective (eds) *But the World Goes on the Same*, Whitley Bay: Erdesdun Publications.
Baggott, R. (1991) 'The politics of public health', *Journal of Social Policy*, 20, 2: 191–214.
Bains, M. (chairman) (1972) *The New Local Authorities: Management and Structure*, London: HMSO.
Barker, T. (1993) 'Is green growth possible', *New Economy*, Sample Issue, Autumn: 20–5.
Bartlett, W. (1991) 'Quasi-markets and contracts: a markets and hierarchies perspective on NHS reform', *Public Money and Management*, Autumn: 53–60.
Bartley, M. (1991) 'Health and labour force participation: 'stress', selection and the reproduction costs of labour power', *Journal of Social Policy*, 18, 1: 327–64.
Batley, R. and Stoker, G. (1991) *Local Government in Europe*, Basingstoke: Macmillan.
Bauman, Z. (1987) *Legislators and Interpreters*, Cambridge: Polity Press.
Beavis, S. and Nicholson, J. (1993) 'Rise and rise of the quangocrats', *The Guardian*, 19 November: 18.
Benington, J. (1975) *Local Government Becomes Big Business*, London: Community Development Project.
Bennett, R. J. (1994) 'Training and Enterprise Councils: are they cost-efficient?' *Policy Studies*, 15, 1: 42–55.
Bennett, R. J., Wicks, P. and McCoshan, A. (1994) *Local Employment and Business Services: Britain's Experiment with TECs*, London: UCL Press.
Bishop, K. (1991) 'Come on down, the price is right in the country', *Planning*, 939: 14–15.
Blackman, T. (1988) 'Housing policy and community action: a comparative study', unpublished Ph.D. thesis, University of Durham.
Blackman, T. (1991) *Planning Belfast: a case study of public policy and community action*, London: Avebury.
Blackman, T. (1992) 'Improving quality through research', in I. Sanderson (ed.) *Management of Quality in Local Government*, Harlow: Longman.
Blackman, T. (1993a) 'Research in local government: an assessment of qualitative research', *Local Government Studies*, 19, 2: 242–63.
Blackman, T. (1993b) 'The appropriateness of geographical data to the assessment of need and the planning of services', in T. Waterston (ed.) *Child Health and Disadvantage*, Andover: Intercept.
Blackman, T., Harrington, B. and Keenan, P. (1993) 'Housing and health: how

allocations can fight illness and despair', *Housing*, 29, 6: 33–4.

Blackman, T. and Stephens, C. (1993) 'The internal market in local government', *Public Money and Management*, 13, 4: 37–44.

Bloch, A. (1992) *The Turnover of Local Councillors*, York: Joseph Rowntree Foundation.

Bongers, P. (1990) *Local Government and 1992*, Harlow: Longman.

Bongers, P. (1992) *Local Government in the Single European Market*, Harlow: Longman.

Bradford, M. G. and Steward, A. (1988) *Inner city refurbishment: an evaluation of private–public partnership schemes*, Centre for Urban Policy Studies Working Paper 3, Manchester: University of Manchester.

Bradshaw, J. (1972) 'The concept of need', *New Society*, 30 March.

Bradshaw, J. (1992) 'The conceptualisation and measurement of need: a social policy perspective', Unpublished paper given at the ESRC seminar Social Research and Health Needs Assessment, University of Salford, January.

Bramley, G. and Le Grand, J. (1992) *Who Uses Local Services? – Striving for Equity*, The Belgrave Papers 4, London: Local Government Management Board.

Brent Council, Brent and Harrow Commissioning Agency and Brent and Harrow Family Health Services Authority (1993) *Partners for Health: A joint health strategy for the London Borough of Brent*.

Brimacombe, M. (1991) 'Taking stock of Estate Action', *Housing*, February: 25–9.

Broady, M. and Hedley, R. (1989) *Working Partnerships: Community development in local authorities*, London: National Council for Voluntary Organisations.

Brooke, R. (1990) *The Environmental Role of Local Government*, Luton: Local Government Training Board.

Brooke, R. (1992) 'Local versus central government – can the tensions be resolved?', *Town & Country Planning*, June: 178–81.

Brooke, R. (1994) 'Radical reorganisation at Tower Hamlets', *AMA News*, April: 22.

Brotchie, J. and Hills, D. (1991) *Equal Shares in Caring*, London: Socialist Health Association.

Bulmer, M., Sykes, W., McKennell, A. and Bailey, C. S. (1993) 'The Profession of Social Research', in W. Sykes, M. Bulmer and M. Schwerzel (eds) *Directory of Social Research Organisations in the United Kingdom*, London: Mansell.

Byrne, D. S., Harrisson, S. P., Keithley, J. and McCarthy, P. (1986) *Housing and Health*, Aldershot: Gower.

Byrne, D. (1989) *Beyond the Inner City*, Milton Keynes: Open University Press.

Byrne, D., McCarthy, P., Keithley, J. and Harrison, S. (1985) 'Housing, class and health: an example of an attempt at doing socialist research', *Critical Social Policy*, 13: 49–72.

Canter, H. (1993) 'Government social research', in W. Sykes, M.Bulmer and M. Schwerzel (eds) *Directory of Social Research Organisations in the United Kingdom*, London: Mansell.

Carter, N., Brown, T. and Abbott, T. (1992) 'The rejuvenation of policy planning?', *Local Government Policy Making*, 19, 2: 44–9.

Castells, M. (1979) *City, Class and Power*, London: Macmillan.

Central Statistical Office (1993) *Social Trends* 23, London: HMSO.

Clarke, M. and Hasdell, I. (1992) 'Toppling the ivory towers', *Municipal Journal*, 44.

Cm 289 (1988) *Public Health in England*, London: HMSO.

Cm 2250 (1993) *Realising our Potential: A Strategy for Science, Engineering and Technology*, London: HMSO.

Cockburn, C. (1977) *The Local State*, London: Pluto Press.

Cole, O. and Farries, J. S. (1986) 'Rehousing on medical grounds – assessments of its effectiveness', *Public Health*, 100: 229–35.

Community Development Project (1974) *Inter-Project Report*, London: CDP Information and Intelligence Unit.

Community Development Project (1975) *Forward Plan 1975–76*, London: CDP Information and Intelligence Unit.

Cooke, B. (1992) 'Quality, culture and local government', in I. Sanderson (ed.) *Management of Quality in Local Government*, Longman, Harlow.

Cooke, P. (1993) 'European experiences of regional economic development', in A. Roberts (ed.) *Power to the People? Economic self-determination and the regions*, London: European Dialogue and the Friedrich Ebert Foundation.

Coombes, M. (1993) 'Community boundaries and commuting areas', in A. Atkinson, T. Blackman and M. Howett (eds) *Research for Policy: Proceedings of the 1993 Annual Conference of the Local Authorities Research and Intelligence Association*, Newcastle upon Tyne: Newcastle City Council.

Coombes, M., Openshaw, S., Wong, C., Raybould, S., Hough, H. and Charlton, M. E. (1992) *Application of Geographic Information Systems to community boundary definition*, London: Department of the Environment, London.

Coopers and Lybrand/Business in the Community (1992) *Growing business in the UK – lessons from continental Europe: Promoting partnership for local economic development and business support in the UK*, London: Coopers and Lybrand.

Cross, A. and Openshaw, S. (1991) 'Crime pattern analysis: the development of ARC/CRIME', Paper presented to the Third National Conference of the Association of Geographic Information, Birmingham, 20–22 November.

Cruddas Park Community Trust (1991) *Annual Report 1990/91*, Newcastle upon Tyne: Cruddas Park Community Trust.

Curtis, S. (1989) *The Geography of Public Welfare Provision*, London: Routledge.

Dale, A. (1992) 'Issues in using the 1991 census', in A. Atkinson, T. Blackman and M. Howett (eds) *Research for Policy: Proceedings of the 1993 Annual Conference of the Local Authorities Research and Intelligence Association*, Newcastle upon Tyne: Newcastle City Council.

Danson, M. W. and Lloyd, M. G. (1992) 'The erosion of a strategic approach to planning and economic regeneration in Scotland', *Local Government Policy Making*, 19, 1: 46–54.

Davies, J. G. (1972) *The Evangelistic Bureaucrat*, London: Tavistock.

Davies, W. K. D. and Herbert, D. T. (1993) *Communities within cities: an urban social geography*, London: Belhaven.

Dawson, H. (1992) 'On the front line', *Local Government Management*, 1, 3: 12–14.

de Groot, L. (1992) 'City Challenge: competing in the urban regeneration game', *Local Economy*, 7, 3: 196–209.

Delamothe, E. (1991) 'Social inequalities in health', *British Medical Journal*, 303: 1046–50.

Delderfield, J., Puffitt, R. and Watts, G. (1991) *Business Planning in Local Government*, Harlow: Longman.

Dennis, N. (1972) *Public Participation and Planners' Blight*,London: Faber & Faber.

Department for Education (1993a) 'A common funding formula for self-governing (grant-maintained) state schools', Department for Education paper, 10 June, London: Department for Education.

Department for Education (1993b) *Technology Colleges: Schools for the Future*, London: Department for Education.

Department for Education (1994) *A common funding formula for self governing (GM) schools 1995–96*, London: Department for Education.

Department of the Environment (1977) *Inner Area Studies: Liverpool, Birmingham*

and Lambeth, London: HMSO.
Department of the Environment (1990) *This Common Inheritance*, London: HMSO.
Department of the Environment (1991) *City Challenge: A New Approach for Inner Cities*, London: Department of the Environment.
Department of the Environment (1992) *Policy Guidance to the Local Government Commission for England*, London: Department of the Environment.
Department of the Environment (1993a) 'Local Government Finance Settlement Working Group Standard Spending Assessments Subgroup – Further Analysis of the Other Services Block (OSB): Testing the New Composite Indices in the Other Services Block', Unpublished paper, Department of the Environment, London.
Department of the Environment (1993b) *Report of the Department of the Environment Joint Working Party on Internal Management*, London: Department of the Environment.
Department of the Environment (1994) *The Local Government Finance Report (England) 1994/5*, London: HMSO.
Department of the Environment (1994a) *SSA Handbook 1994–95*, London: Department of the Environment.
Department of Health (1991) *Research for Health: A research and development strategy for the NHS*, London: Department of Health.
Department of Health (1992) *The Health of the Nation*, London: HMSO.
Department of Health (1993a) *Implementing Community Care: Population Needs Assessment Good Practice Guidance*, London: Department of Health.
Department of Health (1993b) *The Health of the Nation Key Area Handbook: Mental Illness*, London: Department of Health.
Department of Health (1993c) *The Health of the Nation Key Area Handbook: HIV/AIDS and Sexual Health*, London: Department of Health.
Department of Health and Social Security (1980) *Inequalities in Health*, Report of a Research Working Group chaired by Sir Douglas Black, London: DHSS.
Departments of Health and Social Security (1993) *Government Response to the Third Report from the Health Committee Session 1992–93. Community Care: Funding from April 1993*, London: HMSO.
Dobson, A. (1992) 'Using Family Health Service Authority records as a basis for local population stock estimates', in T. Blackman (ed.) *Research for Policy: Proceedings of the 1992 Annual Conference of the Local Authorities Research and Intelligence Association*, Newcastle upon Tyne: Newcastle City Council.
Doern, G. B. (1993) 'The UK Citizen's Charter: origins and implementation in three agencies', *Policy and Politics*, 21, 1: 17–29.
Donnison, D. *et al.* (1991) *Urban Poverty, the Economy and Public Policy*, Combat Poverty Agency, Dublin.
Donnison, D. and Middleton, A. (1987) *Regenerating the Inner City: Glasgow's Experience*, London: Routledge and Kegan Paul.
Doyal, L. and Gough, I. (1991) *A Theory of Human Need*, London: Macmillan.
Dryzek, J. S. (1990) *Discursive Democracy*, Cambridge: Cambridge University Press.
Duncan, S. and Goodwin, M. (1988) *The Local State and Uneven Development*, Cambridge: Polity Press.
Dunleavy, P. (1980) *Urban Political Analysis*, London: Macmillan.
Dunleavy, P. (1981) *The Politics of Mass Housing in Britain*, Oxford: Clarendon Press.
Dunleavy, P. and O'Leary, B. (1987) *Theories of the State: The Politics of Liberal Democracy*, Basingstoke: Macmillan Education.
Durham County Council (1991) *Caring for People in County Durham: The County Council's Community Care Plan 1992/93*, Durham: Durham County Council

Social Services Department.

Easteal, M. (1993) 'Right for Britain', *Municipal Review and AMA News*, 63, 735: 259.

Ellis, R. (1990) 'A British Standard for University Teaching', Unpublished paper, Faculty of Health and Social Sciences, University of Ulster, Northern Ireland.

Emmerich, M. and Peck, J. (1992) 'Strategy Versus Short-Termism in the New Training System: A Tale of Two TECs', *Local Government Policy Making*, 19, 1: 18–25.

Emms, P. (1990) *Social Housing: A European dilemma?*, Bristol: School for Advanced Urban Studies, University of Bristol.

Employment Department Group, Department of Education and Science and Welsh Office (1991) *Education and Training for the 21st Century*, Cm 1536.

Employment Service (1991) *The Jobseeker's Charter*, London: Employment Service.

Fairclough, T. (1992) 'An analysis in need of some improvement', *Public Finance and Accountancy*, 28 August: 12–13.

Fenton, M. (1992) 'Economic development strategies: preparation in a time of retrenchment', *Local Government Policy Making*, 18, 5: 21–23.

FEU (1993) *Value Added in Further Education*, FEU Bulletin, October.

Finkelstein, V. (1993) 'Disability: a social challenge or an administrative responsibility?', in J. Swain, V. Finkelstein, S. French and M. Oliver (eds) *Disabling Barriers – Enabling Environments*, London: Sage.

Forrest, R. and Murie, A. (1990) *Residualisation and Council Housing: A Statistical Update*, Working Paper 91, Bristol: School for Advanced Urban Studies, University of Bristol

Forsyth, L. (ed.) (1988) 'User-controlled community technical aid: a symposium', *Town Planning Review*, 59, 1: 1–44.

French, W. L. and Bell, C. H. (1984) *Organisation development: behavioral science interventions for organisation improvement*, New Jersey: Prentice-Hall.

Friends of the Earth (1989) *The Environmental Charter for Local Government*, London: Friends of the Earth.

Gallant, V. (1992) 'Community involvement: North Tyneside's approach', *Local Government Policy Making*, 18, 4: 42–5.

Gardner, C. and Sheppard, J. (1989) *Consuming Passion: The Rise of Retail Culture*, London: Unwin Hyman.

Gaster, L. (1991) 'Quality and decentralisation: are they connected?' *Policy and Politics*, 19, 4: 257–67.

Gaster, L. (1992) 'Quality in service delivery: competition for resources or more effective use of resources?', *Local Government Policy Making*, 19, 1: 55–64.

Geddes, M. and Benington, J. (eds) (1992) *Restructuring the Local Economy*, Harlow: Longman.

Georghiou, C. (1991) 'Consultation key as city faces up to challenge', *Planning*, 935: 18–19.

Gibbons, J. (1991) 'Childen in Need and their Families: Outcomes of Referral to Social Services', *British Journal of Social Work*, 21: 217–27.

Gibson, T. (1993) 'Walking through treacle', *ARVAC Bulletin*, 52, Spring: 12–13.

Giddens, A. (1984) *The Constitution of Society*, Cambridge: Polity Press.

Glasgow City Council (1992) *Housing Plan for the 90s*, Glasgow: Glasgow City Council.

Gosschalk, B. and Page, B. (1993) 'The new Chartists', *Municipal Journal*, 25 November–2 December, 47: 20–1.

Gostick, C. (1993) 'Social research in local government', in W. Sykes, M. Bulmer and M. Schwerzel (eds) *Directory of Social Research Organisations in the United*

Kingdom, London: Mansell.
Government of the Netherlands Ministry of Housing, Physical Planning and Environment (1989) *To Choose or to Lose: National Environment Plan*, The Hague: Government of the Netherlands.
Graham, H. (1984) *Women, Health and Family*, Brighton: Wheatsheaf Books.
Granville, D. (1993) 'Brum goes for quality services', *Local Government News*, June: 18, 27.
Green, A. and Steedman, H. (1993) *Educational Provision, Educational Attainment and the Needs of Industry: A Review of Research for Germany, France, Japan, the USA and Britain*, London: National Institute of Economic and Social Research.
Grieve, J. and Currie, E. (1990) *Homeless in Britain*, Housing Research Findings 10, York: Joseph Rowntree Memorial Trust.
Gyford, J. (1991) *Does Place Matter? – Locality and Local Democracy*, The Belgrave Papers 3, Luton: Local Government Management Board.
Gyford, J. (1993) 'Rediscovering the local: does place matter?', in M. Howett, A. Atkinson and T. Blackman (eds) *Research for Policy: Information strategy, the 1991 census, place and community*, Proceedings of the 1993 Annual Conference of the Local Authorities Research and Intelligence Association, Newcastle: Newcastle City Council.
Habermas, J. (1984) *The Theory of Communicative Action I: Reason and the Rationalization of Society*, Boston: Beacon Press.
Hall, N. (1990) 'Health: Planning's Forgotten Purpose', *The Planner*, 21 September: 13–16.
Hambleton, R. (1992) *Rethinking Management in Local Government*, Papers in Planning Research 130, Cardiff: Department of City and Regional Planning, University of Wales College of Cardiff.
Hambleton, R. and Hoggett, P. (1990) *Beyond Excellence – Quality Local Government in the 1990s*, Working Paper 85, Bristol: University of Bristol: School for Advanced Urban Studies.
Hamer, M. (1994) 'Change of heart on roads', *New Scientist*, 141, 1911: 7.
Hancox, A., Worrall, L. and Pay, J. (1989) 'Developing a customer-oriented approach to service delivery: the Wrekin approach', *Local Government Studies*, 15, 1: 16–24.
Hannah, J. (1991) 'Transport policy, planning and accident prevention' *HFA 2000 News*, 14: 1–5.
Hansard (1991) 'Local Government Review (England)' cols. 401–30, 21 March, London: HMSO.
Hansard (1992a) *Written Answers*, 13 July, London: HMSO.
Hansard (1992b) *Written Answers*, 15 July, London: HMSO.
Hansard (1992c) *Written Answers*, 16 July, London: HMSO.
Hansard (1992d) 'Recycling debate', 19 June 1992, London: HMSO.
Hansard (1992e) 'Local Government Finance' cols. 136–228, 4 February.
Hansard (1993) *Written Answers*, 17 December, London: HMSO.
Harvey, D. (1981) 'The urban process under capitalism: a framework for analysis', in M. Dear and A. J. Scott (eds) *Urbanization and Urban Planning in Capitalist Society*, London: Methuen.
Harvey, D. (1989) *The Urban Experience*, Oxford: Blackwell.
Hasluck, C. and Duffy, K. (1992) 'Explaining the operation of local labour markets', in M. Campbell and K. Duffy (eds) *Local Labour Markets: Problems and Policies*, Harlow: Longman.
Haughton, G. (1991) 'In search of a moving target – skills surveys and skills audits', *Local Economy*, 6, 2: 177–83.

Haughton, G. and Peck, J. (1988) 'Skills audits: a framework for local economic development', *Local Economy*, 3, 1: 11–20.

Hayton, K. (1989) 'Getting people into jobs', *Local Economy*, 3,4: 279–93.

Healey, P. (1982) 'Understanding land use planning', in P. Healey, J. McDougall, G. Pay and M. J. Thomas (eds) *Planning Theory: Prospects for the 1980s*, Oxford: Pergamon Press.

Heath, T. (1993) 'Placemen take control of Quangoland', *The Guardian*, 19 November: 19.

Hedges, A. and Kelly, J. (1992) *Identification with local areas: summary report on a qualitative study*, London: Department of the Environment.

Henderson, P. (1991) *Signposts to Community Economic Development*, London: Community Development Foundation in association with the Centre for Local Economic Strategies.

Henley, D. F. (1991) *Health in Gateshead 1990/91: Annual Report of the Director of Public Health*, Gateshead: Gateshead Health Authority.

Hepworth, M. (1990) 'Planning for the Information City: the UK Case', Unpublished summary of survey findings, Centre for Urban and Regional Development Studies, University of Newcastle upon Tyne.

Her Majesty's Stationery Office (HMSO) (1977) *Policy for the Inner Cities*, Cmnd 6845, London: HMSO.

Hill, M., Blackman, T., Wallace, B. and Woods, R. (1991) *The Walker Riverside Study: report and recommendations*, Newcastle upon Tyne: Department of Social Policy, University of Newcastle Upon Tyne.

Hirsch, D. (1989) 'Customer-Friendly Council Housing', *Search*, 3:20–3.

Hogg, C. (1991) *Healthy Change*, London: Socialist Health Association.

Hoggett, P. (1991) 'A new management in the public sector?', *Policy and Politics*, 19, 4: 243–56.

Holman, B. (1993) 'Poor lore', *The Guardian*, 15 September.

Holmes, A. (1991) *Limbering Up: Community empowerment on peripheral estates*, Middlesbrough: Radical Improvements for Peripheral Estates Ltd.

Holmes, M. (1993) 'Buying green', *Municipal Journal*, 26 November–2 December, 47: 26, 28.

House of Commons Employment Committee (1988) *Third Report: The Employment Effects of the Urban Development Corporations*, London: HMSO.

House of Lords Select Committee on Science and Technology (1988) *Priorities in Medical Research*, 1987–8 session, London: HMSO.

Hughes, R. (1990) 'Committed change', *Local Government Chronicle*, 14 September: 24.

Hunter, D. J. and Long, A. F. (1993) 'Health research', in W. Sykes, M. Bulmer and M. Schwerzel (eds.) *Directory of Social Research Organisations in the United Kingdom*, London: Mansell.

Husbands, C. (1993) Text of contribution to the conference New Ways of Working with Community Groups, Leeds Metropolitan University, 6 July.

Huxley, P., Hagan, T., Hennelly, R. and Hunt, J. (1990) *Effective Community Mental Health Services*, Aldershot: Avebury.

Illich, I. *et al.* (1977) *Disabling Professions*, London: Marion Boyars.

Inner Cities Grants Division, Department of the Environment (1992) *City Challenge: Bidding Guidance 1993–94*, London: Department of the Environment.

Itzin, C. (1992) 'Hobbled horses', *Local Government Management* Summer: 18–20.

Jackson, P. M. and Palmer, B. (1992) *Developing PerformanceMonitoring in Public Sector Organisations*, Leicester: University of Leicester Management Centre.

Jacobs, M. (1990) *Sustainable development: greening the economy*, London: The Fabian Society.

Jacobs, M. and Stott, M. (1992) 'Sustainable development and the local economy', *Local Economy*, 7, 3: 261–72.

Jenkins, R. (1991) *Food for Wealth or Health*, London: Socialist Health Association.

John, P. and Bloch, A. (1991) *Attitudes to Local Government: A survey of electors*, York: Joseph Rowntree Foundation.

Jones, J. V. (1993) 'Never a golden age, no greener grass', *Public Finance and Accountancy*, 11 June: 39, 41–2.

Jones, G. and Stewart, J. (1994) 'Fewer hands make heavy work', *Local Government Chronicle*, 11 March: 8.

Kelly, J. R. (1992) 'Private sector and public sector views of customer care and quality', in T. Blackman (ed.) *Research for Policy: Proceedings of the 1992 Annual Conference of the Local Authorities Research and Intelligence Association*, Newcastle upon Tyne: Newcastle City Council.

Kelly, P. (1993) 'Waging war on landlords', *Inside Housing*, 11, 25: 8–9.

Kemp, P. (1992) 'Poverty Trap', *Inside Housing*, 9, 46: 9.

Lash, S. and Urry, J. (1987) *The End of Organized Capitalism*, Cambridge: Polity Press.

Lautman, P. and Stearn, J. (1992) 'Distribution of Resources', *Inside Housing*, 9, 40: 10–11.

Lawless, P. (1979) *Urban Deprivation and Government Initiative*, London: Faber & Faber.

Lees, R. and Mayo, M. (1984) *Community Action for Change*, London: Routledge.

Lewis, J. (1992) 'TECs and the provision of training for women', *Local Government Policy Making*, 19, 1: 26–37.

Lewisham Environmental Services (1991) *It's A Deal: Introducing Customer Contracts*, London Borough of Lewisham.

Ley, D. (1983) *A Social Geography of the City*, New York: Harper & Row.

Local Government Information Unit (1992a) *Local Government in the 1990s: A discussion document on the role and functions of local authorities*, London: LGIU.

Local Government Management Board (1992b) *Citizens and Local Democracy: Charting a new relationship*, London: LGMB.

Local Government Management Board (n.d.) *Earth Summit: Rio '92: Information Pack for Local Authorities*, London: Local Government Management Board.

Lowe, C. (1992) 'Education and Training for the 21st Century', *Local Government Policy Making*, 19, 1: 3-8.

Lowndes, V. (1991) 'Decentralisation: the potential and the pitfalls', *Local Government Policy Making*, 18, 2: 19–29.

Lyon, A. (1993) 'Glasgow: a city health plan – more than just a document', *Network News*, UK Health for All Network, Summer.

McArthur, A. (1987) 'Local Economic Initiatives: Community Business', in T. Blackman (ed.) *Community-Based Planning and Development*, Belfast: Policy Research Institute.

McArthur, A. (1993) 'An exploration of community business failure', *Policy and Politics*, 21, 3: 219–30.

McBride, P. (1993) 'Reservations about TECs', *Local Government Policy Making*, 20, 1: 38–43.

McConnell, A. (1990) 'The birth of the poll tax', *Critical Social Policy*, 28: 67–78.

McConnell, C. (1991) *Promoting Community Development in Europe*, London: Community Development Foundation.

Macintyre, S., MacIver, S. and Sooman, A. (1993) 'Area, class and health: should we be focusing on places or people?' *Journal of Social Policy*, 22, 2: 213–34.

McLeod, J. (1992) 'Short life forecast for the council tax', *Public Finance and Accountancy*, 17 July: 10.

Macnicol, J. (1987) 'In pursuit of the underclass', *Journal of Social Policy*, 16, 3: 293–318.

McPhee, D. (1993) 'Tyne sides', *The Guardian*, September 8: 12–13.

McPherson, A. (1992) *Measuring Added Value in Schools*, National Commission on Education Briefing Paper 1.

March, J. (1994) 'A mission to be inconsistent', *Municipal Journal*, 28 January–3 February, 4: 14–15.

Marsh, C., Skinner, C, Arber, S., Penhale, B., Openshaw, S. Hobcraft, J., Lievesley, D. and Walford, N. (1991) 'Samples of Anonymized Records', *Journal of the Royal Statistical Society*, Series A, 154, 2: 305–40.

Martin, P. (1994) 'Spending by the rules', *Local Government Chronicle*, 14 January: 18–19.

Marvin, S. (1992a) 'Levels of access to infrastructure networks', in P. Healey *et al.*, *Newcastle's West End: Monitoring the City Challenge Initiative – A Baseline Report*, Newcastle: University of Newcastle upon Tyne.

Marvin, S. (1992b) 'Urban policy and infrastructure networks', *Local Economy*, 7, 3: 225–47.

Maud Report (1967) *Report of the Committee on Management of Local Government*, London: HMSO.

Means, R. (1992) 'The future of community care and older people in the 1990s', *Local Government Policy Making*, 18, 5: 11–16.

Miliband, D. (1992) 'Introduction: expansion and reform', in D. Finegold, E. Keep, D. Miliband, D. Robertson, K. Sisson and J. Ziman, *Higher Education: Expansion and Reform*, London: Institute for Public Policy Research.

Miller, N. and Munn-Giddings, C. (1993) *Whose need is it anyway?*, Social Services Research Group.

Moore, R. (1987) 'The development and role of standards for the older housing stock', Paper given at the Unhealthy Housing: Prevention and Remedies Conference, University of Warwick, December.

Morris, B. (1993) 'Education Acts: a blunted instrument' *Local Government Policy Making*, 19, 5: 25–30.

National Commission on Education (1993) *Learning to Succeed: Report of the Paul Hamlyn Foundation National Commission on Education*, London: Heinemann.

National Council for Voluntary Organisations (1993) *City Challenge: Involving Local Communities*, London: NCVO Publications.

Newcastle City Council (1986) *The Impact of Poll Tax on Newcastle Residents: Ward Analysis*, Newcastle upon Tyne: Policy Services Unit, Newcastle City Council.

Newcastle City Council (1989) *Environmental Audit: Stage 1 The Objectives and the Indicators*, Newcastle upon Tyne: Newcastle City Council.

Newcastle City Council (1991) *Countryside Strategy*, Newcastle: Newcastle City Council.

Newcastle City Council (1992a) 'West Central Route Environmental Statement', Report by the Director of Engineering, Highways and Protection to Environment and Highways Committee, 8 October.

Newcastle City Council (1992b) 'Unitary Development Plan – Review of Consultation and Next Steps', Report by Director of Development to Policy and Resources Committee, 26 February.

Newcastle City Council (1992c) 'Poverty Impact Analysis', Report by Urban

Initiatives Manager to Anti-poverty Advisory Sub Committee, 8 December.

Newcastle City Council (1993a) *Newcastle upon Tyne Budget 1993–1994*, Newcastle: Newcastle City Council.

Newcastle City Council (1993b) 'Review of Standard Spending Assessments', Report to Policy and Resources Committee by the Chief Executive and City Treasurer, 28 July.

Newcastle City Council Social Services Department (1993) 'Allocation of Home Care Resources', Report of Director of Social Services to Policy Sub-Committee, 19 April.

Newman, P. and Kenworthy, J. (1989) *Cities and Automobile Dependence: A Sourcebook*, Aldershot: Gower.

NHS Management Executive (1991) *Purchasing Intelligence*, EL(91)124, October.

NHS Management Executive (1993) *Public Health: responsibilities of the NHS and the roles of others*, HSG(93)56, November.

NIHE (Northern Ireland Housing Executive) (1992) *Standards of Service*, Belfast: NIHE.

Norris, G. (1989) *The Organisation of the Central Policy Capability in Multi-Functional Public Authorities*, Local Authority Management Unit Discussion Paper 89/1, Newcastle upon Tyne: Department of Economics and Government, Newcastle upon Tyne Polytechnic.

Norton, A. (1991) 'Western European Local Government in Comparative Perspective', in R. Batley and G. Stoker (eds), *Local Government in Europe*, London: Macmillan.

Nuyens, Y. (1991) *From Research to Decision Making: Case studies on the use of health systems research*, Geneva: World Health Organisation.

O'Neill, N. (1992) 'European Regions after Maastricht', *Local Government Policy Making*, 19, 3: 20–8.

Organisation for Economic Co-operation and Development (1992) *Urban Policies for Ageing Populations*, Paris: OECD.

Osmond, J. (1993) 'A Welsh perspective', in A. Roberts (ed.) *Power to the People? Economic self-determination and the regions*, London: European Dialogue and the Friedrich Ebert Foundation.

Page, D. (1993) *Building for Communities: A study of new housing association estates*, York: Joseph Rowntree Foundation.

Pahl, R. E. (1970) *Whose City?*, London: Longman.

Painter, J. (1991) 'Regulation Theory and Local Government', *Local Government Studies*, 17, 6: 23–44.

Pearce, F. (1994) 'A greyer shade of green', *New Scientist*, 141, 1911: 6–7.

Peck, J. and Emmerich, M. (1993) 'Training and Enterprise Councils: Time for Change', *Local Economy*, 8, 1: 4–21.

Pedlar, M., Banfield, P., Boraston, I., Gill, J. and Shipton, J. (1990) *The Community Development Initiative: A Story of the Manor Employment Project in Sheffield*, Farnborough: Avebury.

Percy-Smith, J. (1992) 'Auditing social needs', *Policy and Politics*, 20, 1: 29–34.

Peters, T. J. and Waterman, R. H. (1982) *In Search of Excellence: Lessons from America's best run companies*, New York: Harper & Row.

Petrie, P. (1990) 'School-age child care and local government in the nineteen nineties', *Local Government Policy Making*, 17, 3: 6–12.

PIEDA plc (1992) *Evaluating the Effectiveness of Land Use Planning*, London: HMSO.

Quick, A. (1991) *Unequal Risks: Accidents and Social Policy*, London: Socialist Health Association.

Quick, A. and Wilkinson, R. (1991) *Income and Health*, London: Socialist Health Association.

Raemaekers, J. (1993) 'Corporate environmental management in local government', *Planning Practice and Research*, 8, 3: 5–13.

Rawcliffe, P. and Roberts, J. (1991) 'The art of shortening trip lengths', *Town & Country Planning*, November: 310–11.

Reade, E. (1987) *British Town and Country Planning*, Milton Keynes: Open University Press.

Rees, J. (1988) 'Social polarisation in shopping patterns: an example from Swansea', *Planning Practice & Research*, 6, Winter: 5–12.

Rennie, J. (1993) 'Who is normal?', *Scientific American* 269, 2: 8–10.

Rhodes, R. (1991) 'Introduction: the new public management', *Public Administration*, 69: 1.

Ritchie, C. (1992) 'An exercise in community consultation – devising general criteria', *LARIA News*, 38: 17–19.

Ritchie, C. (1993) 'Information for the Community?' in M. Howett, A. Atkinson and T. Blackman (eds), *Research for Policy: Information strategy, the 1991 census, place and community*, Proceedings of the 1993 Annual Conference of the Local Authorities Research and Intelligence Association, Newcastle: Newcastle City Council.

Roberts, J. (1991) 'Saturday in the supermarket car park – a blip in history?', *Town and Country Planning*, July/August: 204–5.

Robinson, F., Lawrence, M. and Shaw, K. (1993) *More than Bricks and Mortar? Tyne & Wear and Teesside Development Corporations: A Mid-Term Report*, Durham: Department of Sociology and Social Policy, University of Durham.

Rodrigues, J. (1992) 'Curtain up on performance', *Local Government Chronicle*, 20 November: 13.

Rogers, S. (1990) *Performance Management in Local Government*, Harlow: Longman.

Rowthorn, B. (1993) 'Saving the welfare state', *New Economy*, Sample Issue, Autumn: 36–40.

Royal Commission on Local Government in England (1969) *Report of the Royal Commission on Local Government in England 1966–1969*, Cmd 4040, London: HMSO.

Runciman, W. G. (1983) *A Treatise on Social Theory – Volume 1: The Methodology of Social Theory*, Cambridge: Cambridge University Press.

Saunders, P. (1986) 'Reflections on the Dual State Thesis: The Argument, Its Origins and Its Critics', in M. Goldsmith and S. Villadsen (eds) *Urban Political Theory and the Management of Fiscal Stress*, Aldershot: Gower.

Selman, P. (1992) 'Information and monitoring requirements for state of the environment reports and audit: some lessons from the Canadian experience', in T. Blackman (ed), *Research for Policy: Proceedings of the 1992 Annual Conference of the Local Authorities Research and Intelligence Association*, Newcastle: Newcastle City Council.

Shaw, K., Fenwick, J. and Foreman, A. (1993), 'Client and Contractor roles in local government: some observations on managing the split', *Local Government Policy Making*, 20, 2: 22–7.

Sheldon, C. (1992) 'Lean But Green', *Local Government News*, June: 28–9.

Simpson, S. (1993) 'Local Population in 1991: How complete was the census?', in A. Atkinson, T. Blackman and M. Howett (eds) *Research for Policy: Proceedings of the 1993 Annual Conference of the Local Authorities Research and Intelligence Association*, Newcastle upon Tyne: Newcastle City Council.

Skelcher, C. (1992), *Managing for Service Quality*, Longman: Harlow.
Smith, S. J., Alexander, A., Hill, S., McGuckin, A. and Walker, C. (1993) *Housing Provision for People with Health Problems and Mobility Difficulties*, Housing Research Findings, 86, York: Joseph Rowntree Foundation.
Solesbury, W. (1993) 'Reframing urban policy', *Policy andPolitics*, 21, 1: 31–8.
Social Information Systems (1991) *Warwickshire Social ServicesDepartment: A plan to develop management information and performance indicators*, Manchester: Social Information Systems.
Sondhi, R. and Salmon, H. (1992) 'Race, racism and local authorities', *Local Government Policy Making*, 18, 5: 3–9.
Stewart, M. (1990) *Urban Policy in Thatcher's England*, Working Paper 90, Bristol: School for Advanced Urban Studies, University of Bristol.
Stewart, M. (1994) 'Between Whitehall and Town Hall: the realignment of urban regeneration policy in England', *Policy and Politics*, 22, 2: 133–45.
Stoker, G. (1991) *The Politics of Local Government*, London: Macmillan.
Stoker, G. and Young, S. (1993) *Cities in the 1990s*, Harlow: Longman.
Sylva, K. and David, T. (1990) '"Quality" education in preschool provision', *Local Government Policy Making*, 17, 3: 61–7.
Tam, H. (1993) *Serving the Public: Customer Management in Local Government*, Longman, Harlow.
Taylor, D. (1992) 'Environmental auditing in Lancashire', in T. Blackman (ed.), *Research for Policy: Proceedings of the 1992 Annual Conference of the Local Authorities Research and Intelligence Association*, Newcastle: Newcastle City Council.
Taylor, P. (1985) *Political Geography*, London: Longman.
Terry, F. (1991) 'Good Health Matters', *Local Government Chronicle* 16 August: 10.
Thomas, I. (1992) 'Preparing plans poses problems', *Municipal Review and AMA News*, 730: 118.
Thomas, S., Goldstein, H. and Nuttall, D. (1993) 'Number-crunching', *The Guardian Education*, 30 November: 3–4.
Titterton, M. (1992) 'Managing threats to welfare: the search for a new paradigm of welfare', *Journal of Social Policy*, 21, 1: 1–24.
Tomaney, J. (1993) 'An English regional perspective', in A. Roberts (ed.) *Power to the People? Economic self-determination and the regions*, London: European Dialogue and the Friedrich Ebert Foundation.
Tomaney, J. and Turnbull, N. (1992) 'Options against decline – the case for regional government', in R. Forbes and J. Tomaney (eds), *Governing Ourselves*, Special Edition of *Northern Labour and Economy*, Newcastle upon Tyne: Trade Union Studies Information Unit.
Travers, T. (1992) 'Taxing Times', *Roof*, Vol. 17, No. 4: 28–9.
TUC (1992) *The Quality Challenge*, London: TUC.
Twelvetrees, A. (1991) *Community Work*, London: Macmillan.
Tyne and Wear Research and Intelligence Unit (1991) *Tyne and Wear Urban Development Area Progress Report 1988–91*, Newcastle upon Tyne: Tyne and Wear Research and Intelligence Unit.
Tyne and Wear Urban Development Corporation (1992) *Annual Report and Financial Statements 1991/1992.*
Tyne Wear 2000 (1987) *A Regional Government for the North of England*, Newcastle: Tyne Wear 2000.
Tysome, T. (1993a) 'Carrot for FE colleges', *The Times Higher Education Supplement*, 16 July: 2.
Tysome, T. (1993b) 'Snapshot of TECs shows blurred edge', *The Times Higher*

Education Supplement, 17 September: 3.
Vines, G. (1989) 'Exercise your mind', *New Scientist*, 18 March.
Vittles, P. (1991) 'The principles and techniques of qualitative research', Unpublished paper given at the Local Authorities Research and Intelligence Association Autumn Workshop, Wakefield, September.
Ward, C. (1984) 'George Orwell and the Politics of Planning', *The Planner*, 70, 1: 4–6
Warren, M. (1991) 'Another day, another debrief: the use and assessment of qualitative research', *Journal of the Market Research Society*, 33: 1.
Weiss, E. B. (1989) *In Fairness to Future Generations: International Law, Common Patrimony and Intergenerational Equity*, Tokyo: The United Nations University.
Welfare, D. (1993) 'Additionality: the key to understanding it', *AMA News*, 743: 182–3.
Whitehead, M. (1987) *The Health Divide*, London: Health Education Council.
Widdicombe Report (1986) *Report of the Committee of Inquiry into the Conduct of Local Authority Business*, Cmnd. 9797, London: HMSO.
Wilkinson, R. G. (1992) 'Income distribution and life expectancy', *British Medical Journal*, 304: 165–8.
Willmott, P. and Hutchison, R. (1992) *Urban Trends 1*, London: Policy Studies Institute.
Wills, J. (1994) 'Whitehall's suspect sums', *Local Government Chronicle*, 4 March: 14–15.
Wilson, R. A. (1993) *Review of the Economy and Employment 1992/3: Occupational Assessment*, Warwick: Warwick University Institute of Employment Research.
Wolman, H. and Goldsmith M. (1992) *Urban Politics and Policy: A Comparative Approach*, Oxford: Blackwell.
Woods, R. (1991) *Pennywell Neighbourhood Centre: Interim evaluation report*, Newcastle upon Tyne: Department of Social Policy, University of Newcastle upon Tyne.
Worrall, L. and Rao, L. (1991) 'The Telford Urban Policy Information Systems Project', in L. Worrall (ed.) *Spatial Analysis and Spatial Policy using Geographic Information Systems*, London: Belhaven Press.
Worthy, A. (1993) 'CCT ruling backs sacked worker', *Unison Local Government*, September: 8.
Yeomans, K. (1993) 'Estate of the Art', *Housing*, October: 40.
York City Council (1991) *Citizen's Charter 1991–92*, York: York City Council.
York City Council (1993) *The York Citizen's Charter*, April 1993–March 1994, York: York City Council.
York City Council Housing Services (1993a) 'Report to Housing Services Committee by Director of Marketing and Communications: 1993 Annual Service Monitor', 15 November, York: York City Council.
York City Council (1994) *The York Citizen's Charter*, April 1994–March 1995, York: York City Council.

INDEX